I0475980

Time's Arrow (?)

Copyright © 2018 Arieh Ben-Naim.

All rights reserved. No part of this book may be reproduced, stored,
or transmitted by any means—whether auditory, graphic, mechanical,
or electronic—without written permission of the author, except in the
case of brief excerpts used in critical articles and reviews. Unauthorized
reproduction of any part of this work is illegal and is punishable by law.

This book is a work of non-fiction. Unless otherwise noted, the author and the publisher
make no explicit guarantees as to the accuracy of the information contained in this book
and in some cases, names of people and places have been altered to protect their privacy.

ISBN: 978-1-4834-8686-4 (sc)
ISBN: 978-1-4834-8685-7 (e)

Because of the dynamic nature of the Internet, any web addresses or links contained in
this book may have changed since publication and may no longer be valid. The views
expressed in this work are solely those of the author and do not necessarily reflect the
views of the publisher, and the publisher hereby disclaims any responsibility for them.

Any people depicted in stock imagery provided by Getty Images are
models, and such images are being used for illustrative purposes only.
Certain stock imagery © Getty Images.

Lulu Publishing Services rev. date: 06/12/2018

Time's Arrow (?)

The Timeless Nature of
Entropy and the Second Law of
Thermodynamics

Arieh Ben-Naim

**This book is dedicated
to all those who believe that
Entropy is related to the "Arrow of Time"**

הֲבֵל הֲבָלִים אָמַר קֹהֶלֶת, הֲבֵל הֲבָלִים, הַכֹּל הָבֶל (קהלת א,ב)

"Meaningless! Meaningless!" says the teacher.
"Utterly meaningless! Everything is meaningless."
Ecclesiastes (1,2)

Contents

List of Abbreviations

ABN   Arieh Ben-Naim

IT   Information Theory

Pr   Probability, or super probability

SMI   Shannon's measure of information

20Q   Twenty questions

Preface

Open any book which deals with "Time," "Theory of Time," "Arrow of Time," "Time's Beginning," or "Time's Ending," and you are likely to find an association of entropy and the Second Law of Thermodynamics with the so-called "Time's Arrow."

The concept of Time's Arrow and its explicit association with the Second Law is attributed to Eddington (Chapter 1). However, the association of the Second Law with *time* can be traced back to both Clausius and Boltzmann, who related the one-way property of time with the Second Law.

More recently, Mackey (1992) dedicated an entire book to "Time's Arrow." The Origins of Thermodynamic Behavior." It should be mentioned that Mackey's book does not discuss the "Thermodynamic Behavior" in general, but only the Second Law of Thermodynamics. In his book's preface, he quotes both Clausius' and Eddington's statements on the Second Law.

On Clausius' formulation of the Second Law, Mackey comments:

"I find that Clausius' formulation is the most transparent."

On Eddington's statement, he comments:

"I believe that Eddington's pronouncement still carries a great deal of truth."

I sharply disagree with both of these comments made by Mackey. This book aims to debunk both Clausius' and Eddington's statements regarding the Second Law of Thermodynamics. It will also be shown that Boltzmann erred in his interpretation of his H-Theorem. More generally we shall show that *entropy has nothing to do with time.*

The title of my book was chosen to contrast with Mackey's book. First, I doubt that the metaphor of "Time's Arrow" has any physical significance. I have discussed my views on this in my book, Ben-Naim (2016a). This explains the question mark posited after "Time's Arrow." Second, my book's subtitle is: "Timeless Nature of Entropy and the Second Law of Thermodynamics." In fact, no one claims that thermodynamics, in general, is associated with Time's Arrow. The only law that was, and still is associated with time, is the Second Law of Thermodynamics. It is fitting therefore in this book to discuss specifically the timeless character of entropy and the Second Law.

In this current book, I will also argue that entropy cannot be defined for the entire universe. This is tantamount to saying that Clausius, who coined the term "entropy" erred in his enunciation of the Second Law as: "Entropy of the Universe always increases." More on the "Entropy of the Universe" may be found in "The Four Laws That Do Not Drive the Universe," Ben-Naim (2017c).

Similarly, the common statement of the Second Law: "Entropy always increases," is meaningless unless one specifies the *system* for which the entropy is discussed. Once the system is specified, its entropy is fixed and does not change with time.

This book is organized in six chapters. The first chapter includes some historical milestones on the association of entropy with time. Chapter 2 presents three different definitions of entropy. These *definitions* are different, but equivalent; they all lead to the same results whenever entropy changes can be calculated. Chapter 3 deals with the derivation of the *entropy function* based on Shannon's measure of entropy. This derivation and the associated *definition* of entropy is relatively new, and not well-known. Chapter 4 presents a few formulations of the Second Law of Thermodynamics. In this chapter, we shall demonstrate that the most general formulation of the Second Law does not mention entropy. Chapter 5 presents a few examples of changes in entropy in some simple processes. Finally, Chapter 6 discusses Boltzmann's H-theorem, the criticisms surrounding it, and some of the mistakes made by Boltzmann and his critics.

The main message of this book is that Clausius, Boltzmann, Eddington and many others, were misled in believing that there is such a quantity called entropy, which always increases with time, and therefore may be identified with the so-called "Arrow of Time."

The main reasons for this mistaken view lie in the following:

1. Confusing Shannon's measure of information with the thermodynamic entropy;

2. Confusing various concepts of reversibility and irreversibility;

3. Confusing reversal of the system to its initial state, with reversal of the value of the entropy of the system;

Thanks to the new definition of entropy based on Shannon's measure of information, all the confusion was clarified and evaporated. The new definition of entropy shows unequivocally that entropy is not a function of time and does not have a "tendency to increase with time." It also follows that the concept of entropy cannot be applied to either living systems, or to the entire universe. In Appendix B we shall also raise some serious doubts regarding the assumption of *local equilibrium,* which is at the heart of the theory of non-equilibrium thermodynamics.

Arieh Ben-Naim

Department of Physical Chemistry
The Hebrew University of Jerusalem
Jerusalem, Israel
Email: ariehbook@gmail.com
URL: ariehbennaim.com

Acknowledgments

I am grateful to Steven Bottomley, Robert Engel, Robert Goldberg, Jose Angel Sordo Gonzalo, Douglas Hemmick, Shannon Hunter, Richard Henchman, Maik Jacob, Mike Rainbolt and Erik Szabo for reading parts or all of the manuscript and offering useful comments.

As always, I am very grateful for the graceful help from my wife, Ruby, and for her unwavering involvement in every stage of the writing, typing, editing, re-editing, and polishing of the book.

Chapter 1. Introduction and some historical milestones

The explicit association of "Time's Arrow" with entropy is attributed to Eddington (1928). However, the conceptual association between entropy and time may be traced back to Clausius' famous enunciation of the Second Law, Clausius (1879)[1]:

"The entropy of the world tends to a maximum"

Similar statements have been made by many others.

Examples: Brillouin (1962):

"The probability has a natural tendency to increase, and so does entropy"

Both parts of the statement are wrong![2] Neither probability (of what?) nor entropy (of what?) has a "natural tendency to increase."

Similarly, the title of Atkins' Chapter on the Second Law [Atkins (2007)] is:

"The Second Law: The increase in entropy."

This is a misleading statement of the Second Law. More detailed criticism of Atkins' book (2007) may be found in Ben-Naim(2017c).

The statement *"entropy always increases,"* always means that entropy always increases *"with time."* As we shall see, statements like *"entropy always increases"* are meaningless; entropy, by itself cannot be assigned a numerical value, therefore one cannot claim that it increases or decreases. The statement "entropy always increases" is as meaningful as "beauty always increases," or "wisdom always increases." As we shall also see, entropy as a *state-function,* is defined only for well-defined thermodynamic systems at equilibrium. Therefore, entropy changes are meaningful only for well-defined thermodynamic processes, occurring in systems for which the entropy is defined; once we specify the system, then its entropy is determined, and does not change with time. [3]

Eddington (1928) is credited for the explicit association of "The law that entropy always increases" with "Time's Arrow," which expresses this "one-way property of time." Quotations from Eddington feature in most popular science books, as well as in some textbooks on thermodynamics. Here are two relevant quotations from Eddington's (1928) book, "The Nature of the Physical World." The first concerns the role of entropy and the Second Law, and the second, introduces the idea of "time's arrow."

1. *"The practical measure of the random element which can increase in the universe but can never decrease is called entropy…*
The law that entropy always increases, holds, I think, the supreme position among the laws of Nature."

2. *"Let us draw an arrow arbitrarily. If as we follow the arrow we find more and more of the random element in the state of the world, then the arrow is pointing towards the future; if the random element decreases the arrow points towards the past.*

This follows at once if our fundamental contention is admitted that the introduction of randomness is the only thing which cannot be undone. I shall use the phrase 'time's arrow' to express this one-way property of time which has no analogue in space".

In the first quotation Eddington reiterates the *unfounded* idea that "entropy always increases." Although it is not explicitly stated, the second quotation alludes to the connection between the Second Law and the Arrow of Time. This is clear from the (erroneous) association of the "random element in the state of the world" with the "arrow pointing towards the future."

In my view it is far from clear that an Arrow of Time exists. These views were previously discussed in Ben-Naim (2015a, 2016a). While it is true that we feel that "time passes," "time runs slowly or quickly," these are subjective psychological feelings. I doubt that the metaphor of "Time's Arrow" has any physical reality.[4] As I have demonstrated earlier in [Ben-Naim (2012, 2016b)], and as I will also do so in the present book, entropy is not associated with randomness, and that it is not true that entropy always increases.

There are many other statements in Eddington's book which are unfounded and misleading. For instance; the claim that entropy is a *subjective* quantity, the concepts of *"entropy-clock,"* and *"entropy-gradient."* Reading through the entire book by Eddington, you will not find a single correct statement on the thermodynamic entropy!

Here is another typical quotation from Rifkin's book (1980), associating *entropy* with the *Arrow of Time*.

"...the second law. It is the irreversible process of dissipation of energy in the world. What does it mean to say, 'The world is running out of time'? Simply this: we experience the passage of time by the

succession of one event after another. And every time an event occurs anywhere in this world energy is expended and the overall entropy is increased. To say the world is running out of time, then, is to say the world is running out of usable energy. In the words of Sir Arthur Eddington, 'Entropy is time's arrow."

It is not!!!

Such statements were driven to the extremes of absurdity by using the *quality sign* "=" in order to express the identity of time's arrow and entropy. In "The Demon and the Quantum," Scully writes (2007):

"The statistical time concept that entropy = time's arrow has deep and fascinating implications. It therefore behooves us to try to understand entropy ever more deeply. Entropy not only explains the arrow of time; it also explains its existence; it is time."

As we shall see in this book, entropy does not explain the "arrow of time." It does not explain its existence, and it is in *fact*, a timeless quantity! Therefore, although such statements may be *deeply fascinating*, they are meaningless and misleading.

Most writers on entropy and the Second Law, adhering to Clausius' statement of the Second Law claim that since the entropy of the universe always increases, it follows that the entropy of the universe must have been low at, or near, the Big Bang. This is known as the Past Hypothesis; a meaningless hypothesis suggested by Albert (2000). Some authors, e.g. Carroll (2010), even went a step further, and used the Past Hypothesis to *explain* everything that happens, including our ability to remember the past but not the future. Carroll dedicates a whole book to endlessly repeating these silly ideas about the "entropy of the early universe" and its relationship with the Arrow of time. I have criticized Carroll's book in more detail in Ben-Naim (2016a, 2018b)

In Chapter 6, we present what is considered to be the most original theoretical discussion of the change in entropy with time. This is the so-called Boltzmann's H-theorem. We shall devote Chapter 6 to critically examine this theorem. Here we quote from a recent book by Davies (1995), who summarized Boltzmann's theorem for the layperson. On page 37 we find:

"He [Boltzmann] discovered a quantity, defined in terms of the motions of the molecules that provided a measure of the degree of chaos in the gas. This quantity, Boltzmann proved, always increases in magnitude as a result of the molecular collision, suggesting it be identified with thermodynamic entropy. If so, Boltzmann's calculation amounted to a derivation of the second law of thermodynamics from Newton's laws."

As we shall see, the truth is that Boltzmann erred in his interpretation of the H-Function, and this error has been propagated in the literature ever since.

A book by Mackey titled "Time's Arrow. The Origins of Thermodynamic Behavior," was published in 1992. Its aim is to derive the Second Law of Thermodynamics from the dynamics of the particles.[5] As we shall see in Chapter 5, the Second Law is basically a *law of probability,* and such a law cannot, in principle, be derived from Newton's dynamics.

Finally, it is appropriate to quote here a paragraph from a very recent book by Müller (2016).

"Eddington attributed the flow of time to the increase in entropy, a measure of disorder in the universe. We now know enormously more about the entropy of the universe than did Eddington in 1928 when he proposed the theory, and I'll argue that Eddington got it backward. The flow of time causes entropy to increase, not the other way around."

There are several points with which I do not agree with. First, entropy is *not* a measure of "disorder" in the universe. It is quite strange that this erroneous interpretation of entropy still appears in a book published in 2016. See also Ben-Naim (2012, 2016b).

Second, no one knows how to define the entropy of the universe. Therefore, the statement about our knowledge of the entropy of the universe either now, or in 1928 is highly misleading. We now know about the entropy of the universe, as much as we know about the wisdom or the stupidity of the universe.

Third, the statement that the flow of time causes entropy to increase, not the other way around, is doubly misleading. Entropy, by itself, cannot be said to increase or decrease. The reason is that entropy is a *state function*, i.e. it is *defined* for a well-defined system at equilibrium. As such, it is not a function of time. The flow of time is *not* the cause of entropy increase! The "flow of time" (if it flows) has nothing to do with entropy increase!

At this point, we pause to discuss, very briefly, the concept of equilibrium state. Experimentally, we know that systems consisting of a huge number of particles can be described by a few parameters. For instance, a gas consisting of N atoms of argon, can be described by its pressure and its temperature. This description is referred to as the *thermodynamic* or *macroscopic state* of the system. Clearly, a macroscopic state does not specify the *microscopic states* of the system. For these, we need to know the locations and velocities of a huge number of particles, $N \approx 10^{23}$.

As we shall see throughout the book, entropy can be defined only for systems at equilibrium states. Time does not feature in thermodynamics, in general, nor in entropy, in particular; this is what I mean by the:

"Timeless Nature of Entropy and the Second Law of Thermodynamics"

We know also that there exist states in which the thermodynamic parameters, say temperature, pressure, or density do not change with time. These states are called equilibrium states.

It should be stressed however, that there is no general definition of equilibrium which applies to all systems. Callen (1985) introduced the existence of the equilibrium state as a postulate. He also emphasized that any definition of an equilibrium state is necessarily circular.

In practice, we find many systems in which the parameters describing the system seem to be unchanged with time. Yet, they are not equilibrium states. But for all our purposes in this book, we can assume that every well-defined system, say having a fixed energy E, volume V, and number of particles N will tend to an equilibrium state. At this state, the entropy of the system, as well as many other thermodynamic relationships are applicable.

Chapter 2. The various definitions of Entropy

This chapter introduces three definitions of entropy.[6] The first is referred to as either the thermodynamic, experimental, macroscopic, or the non-atomistic definition. This definition originated in the 19th century, stemming from the interest in heat engines. The introduction of entropy into the vocabulary of physics is attributed to Clausius. In reality, Clausius did not *define* entropy, but rather only changes in entropy. Clausius' definition, together with the Third Law of Thermodynamics, led to the calculation of "absolute values" of the entropy of many substances.[7]

There are other quantities which are referred to as entropy[8] but have nothing to do with the thermodynamic entropy.

The second definition is attributed to Boltzmann. This definition is sometimes referred to as either the microscopic definition, or the atomistic definition of entropy. It relates the entropy of a system to the number of accessible microstates of a thermodynamic system characterized macroscopically by the total energy E, volume V, and total number of particles N [for a multi-component system N may be reinterpreted as the vector $(N_1, ..., N_c)$, where N_i is the number of atoms of type i]. The extension of Boltzmann's definition to systems characterized by independent variables other than the (E, V, N) is attributed to Gibbs. Gibbs also introduced the idea of an ensemble of systems and showed how one can reformulate the Second Law of Thermodynamics for systems characterized by other sets of independent variables such as T, V, N or T, P, N. It is well known that the Helmholtz energy attains a minimum value for a system characterized by the variables T, V, N. Similarly, the Gibbs energy attains a minimum value for a system characterized by the variables T, P, N. Ultimately all these reformulations of the Second Law can be traced back to the formulation in terms of maximum value of the entropy of an isolated system (i.e. E, V, N constants). As we shall see in Chapter 4, "maximum entropy" does not mean maximum as a function of time, but maximum over all possible constrained equilibrium states[7,8]

The Boltzmann definition, see Section 2.2, seems to be completely unrelated to the Clausius definition, see Section 2.1. However, it is found that for all processes for which entropy changes can be calculated by using Boltzmann's definition, the results agree with entropy changes, calculated using Clausius' definition. Although there is no formal proof that Boltzmann's entropy is equal to the thermodynamic entropy, as defined by Clausius, it is widely believed that this is true.

The third definition which I will refer to as the ABN definition, is based on Shannon's measure of information (SMI). It may also be referred to as the microscopic or the atomistic definition of entropy. However, this definition of the entropy, as well as the Second Law is very different from Boltzmann's definition. This is also the only definition which provides a simple, intuitive, and meaningful interpretation of entropy and the Second Law.[9]

We also note that calculations of entropy changes based on the SMI definition agrees with those calculations based on Clausius', as well as on Boltzmann's definition. Unlike Boltzmann's definition, the SMI definition does not rely on calculations of the number of accessible states of the system. It also provides directly the *entropy function* of an ideal gas, and by extension, also the entropy function for a system of interacting particles.

In each of the following sections we shall examine the question of time dependence of entropy. As we shall see in Chapter 4, the independence of entropy on time is most clearly revealed from the SMI-based definition of entropy. Therefore, we shall devote Chapter 3 to derive the entropy from the SMI.

2.1 The Clausius definition

In this section, we start with the early considerations regarding heat engines which led Clausius to introduce the concept of entropy and formulate the Second Law of Thermodynamics.

2.1.1 Heat engines and Carnot's efficiency

Traditionally, the birth of the Second Law is associated with the name Sadi Carnot (1796-1832). Although Carnot himself did not formulate the Second Law, his work laid the foundations on which the Second Law was formulated a few years later by Clausius and Kelvin. Carnot was interested in heat engines, more specifically, in the *efficiency* of heat engines.

In the 18th century, scientists believed that *heat* was a kind of fluid called *caloric* that flows from a higher to a lower temperature. Today, this "caloric theory" is considered obsolete. We shall still use the term "heat flow" meaning heat transfer.

Qualitatively, think of a waterfall and imagine that for some quantity of water falling from h_2 to h_1 you can do some useful work. Likewise, for a given amount of heat "falling" from the higher temperature T_2 to a lower temperature T_1,[10] you can do some useful work. Figure 2.1.

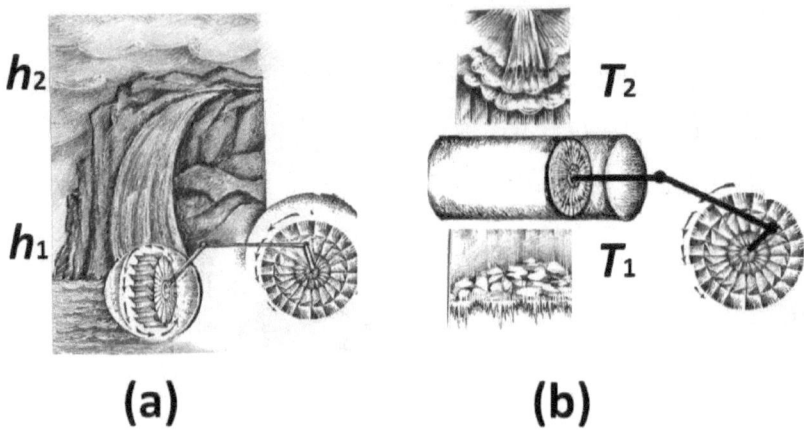

(a)   **(b)**

Figure 2.1. (a) A "water fall," and (b) A "heat engine."
These two processes are not driven by the same law.

Carnot was interested in the *efficiency* of a heat engine, how much useful work one can get from a given amount of heat that flows from the higher temperature T_2 to the lower temperature T_1. Carnot found, somewhat unexpectedly, that there is a limit on the efficiency of a heat engine operating between two temperatures T_2 and T_1.[11] This finding was not a formulation of the Second Law, but it sowed the seeds for the inception of the Second Law.

2.1.2. Clausius' definition of changes in entropy

Basically, Clausius observed, as every one of us does, that there are many processes that occur in nature spontaneously and always in one direction. Examples abound.

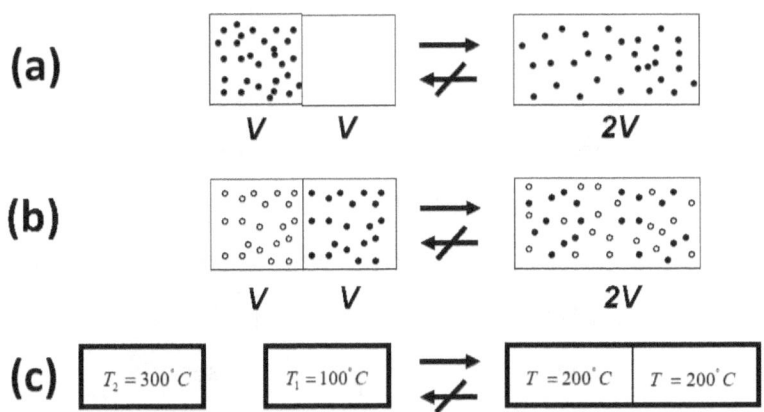

Figure 2.2. Three spontaneous processes occurring in isolated systems typically discussed in connection with the Second Law.
(a) Expansion of an ideal gas from V to 2 V
(b) Mixing of two different ideal gases
(c) Heat transfer from a hot to a cold body

1. Take gas in a small box and open it within a larger empty box. You will *always* observe that the gas will expand and fill the entire volume of the bigger box, Figure 2.2a.

2. Take two gases, say argon and neon in different compartments separated by a partition. Remove the partition, and you will always observe a spontaneous mixing of the two gases, Figure 2.2b.

3. Take two identical pieces of iron, one at 300°C and the second at 100°C. Bring them to thermal contact, Figure 2.2c. You will always observe a spontaneous flow of heat from the body with the higher temperature to the

body with the lower temperature. The hotter body will cool down; the cooler body will heat up. At equilibrium the two bodies will have a uniform temperature of 200°C.

Before we continue it must be said that all these processes are supposed to occur in isolated systems. One can reverse all of these processes by using an external agent.

In all of these examples you *never* observe the reverse process spontaneously; the gas *never* condenses into a smaller region in space, the two gases *never* un-mix spontaneously after being mixed, and heat *never* flows from a cold to a hot body. Note carefully that I italicized the word *never* in the previous sentences. Indeed, we *never* observe any of these processes occurring spontaneously in the reverse direction. For this reason, the processes shown in Figure 2.2 (as well as many others) are said to be *irreversible*. The idea of absolute *irreversibility*, is not only not true, it is also erroneously associated with the very definition of entropy. One should be careful with the use of the words "reversible" and "irreversible" in connection with the Second Law. There are several, very different definitions assigned to these words. For more details, see Appendix A, and Ben-Naim (2011a, 2015a, 2016a). Here, we point out two possible definitions of the term *irreversible*.

1. We *never* observe that the final state of any of the processes in Figure 2.2 returns back to the initial state (on the left-hand side of Figure 2.2) spontaneously.

2. We *never* observe the final state of any of the processes in Figure 2.2 going back to the initial state and *staying* in that state.

In case 1, the word *never* is used in "practice." The system can go from the final to the initial state. In this case, we can say that the initial state will be *visited*. However, such a reversal of the process would occur once in many ages of the universe. Therefore, this is *practically* an irreversible process; we will "never" observe such a reversal in practice.

In case 2, the word *never* is used in an absolute sense. The system will never go back to its initial state and stay there!

In section 4.1 we will discuss a few processes of expansion with small numbers of particles. In such systems when we remove the partition between the two chambers, the particles will move to occupy the larger volume spontaneously. In such systems the particles will move to occupy the larger volume spontaneously when we remove the partition between two chambers. However, from time to time we will observe all the particles in one chamber (either the left, or the right one). However, once they are in one chamber (and this will occur with decreasing probability as the number of particles increase), the particles will expand again to occupy all the available volume. In order to *stay* in one chamber, the partition should be replaced at its original position. Of course, this will never (in an absolute sense) occur spontaneously.

The distinction between these two "irreversibilities" is important in connection with the formulation of the Second Law. Going back to the three processes shown in Figure 2.2, we can ask *why* these processes always occur in one direction. Is there a law of Nature that dictates the direction of the unfolding of these processes? Look again at the three processes depicted in Figure 2.2. Take note that these are quite *different* processes and that it is not clear whether they are all governed by the same law. Perhaps, there is one law for the spontaneous expansion of a gas, another for the spontaneous mixing of two gases, and still another for the spontaneous flow of heat from a hot to a cold body.

It was Clausius who realized the common principle underlying all these processes and postulated that there is only one law that governs all these processes. Even before formulating the Second Law, Clausius' postulate was an outstanding achievement considering the fact that none of these processes were understood. You watched how a colored gas expands and fills a larger volume. You watch a drop of blue ink mixing with a glass of water coloring the entire liquid. You watch a hot body cooling and a cold body heating. You watch all of these with your *macroscopic eyes*, but you have no idea what *drives* these processes, what goes on *inside* the systems you are watching. Such an insight was not even possible before the atomic nature of matter was embraced by the scientific community which allowed us to use our "*microscopic eyes*" to "see" what goes on when such processes occur. "Seeing," even with our microscopic eyes, is one thing, and *understanding* what we see is quite another.

Clausius started from one particular process; the spontaneous flow of heat from a hot to a cold body. Based on this specific process Clausius defined a new quantity which he called Entropy. Let $dQ > 0$ be a *small quantity* of heat flowing into a system, being at a given temperature T. The *change* in *entropy* was defined as [12]

$$dS = \frac{dQ}{T} \qquad\qquad (2.1)$$

The letter d here stands for a very small *quantity*, and T is the absolute temperature. Q has the units of *energy*, and T has the units of *temperature*. Therefore, the entropy change has the units of *energy* divided by units of *temperature*. The quanity of heat, dQ *must* be very small, such that when it is transferred into, or out from the system, the temperature T does not change. If dQ is a finite quantity of heat, and one transfers it to a system which is initially at a given T, the temperature of the system might change, and therefore the change in entropy will depend on both the initial and the final temperature of the system. Note carefully that this equation does not define *entropy* but only changes in entropy for a particular process, i.e. a *small* exchange of heat ($dQ > 0$ means heat flows into the system, $dQ <$ 0 means heat flows out of the system). There are many processes which do not involve heat transfer (e.g. Figure 2.2a and 2.2b. In 2.2c we consider the two parts of the system as one isolated system; within this system there is a flow of heat from one part to the other). Yet, from Clausius' definition, and the postulate that the entropy is a *state function*, one could devise a *path* leading from one state to another, for which the entropy change can be calculated.

It is not uncommon to refer to the equation $dS = dQ/T$ as Clausius' *definition* of entropy. In fact, this equation does not define entropy, nor *changes* in entropy for every process (e.g. expansion of an ideal gas).

At this stage we note that this definition requires the system to be at a fixed temperature. A system at a fixed temperature must be at *thermal* equilibrium but does not have to be at equilibrium. For instance, a chemical reaction might take place within the system at constant temperature.

However, if one wants to calculate finite changes between two macroscopic states of a system one must assume that the system is at equilibrium in both the initial and in the final states of the system.

Initially, Clausius formulated a "restricted" Second Law, namely that heat does not flow spontaneously from a cold to a hot body. However, he later postulated that there exists a quantity, which he called entropy, which is assigned to any macroscopic system, wherein the entropy always increases when a spontaneous process occurs. This was the birth of the Second Law of Thermodynamics. This law introduced a new quantity to the vocabulary of physics, and at the same time brought together many processes under the same umbrella.

The extraordinary achievement of Clausius was the enormous generalization from a few spontaneous processes to *any* spontaneous process. It should be stressed here that the formulation of the Second Law in terms of entropy is valid only for isolated systems, i.e. systems having a constant energy, volume and number of particles. For other systems, say P, T, N the entropy can either go up or down. We shall discuss these formulations of the Second Law in Chapter 5.

Soon after Clausius formulated the Second Law scientists proved that various particular formulations were all equivalent to each other. The proofs of the equivalency appear in any textbook of thermodynamics. Today, we can calculate the change in entropy, and we find that whenever a spontaneous process occurs in an *isolated* system, the entropy of the system always increases, or remains unchanged. We shall discuss the various formulations of the Second Law in Chapter 5. Before we continue we must emphasize that by "entropy changes" we mean difference in entropy between two *equilibrium states*. We do not know how to calculate the entropy change for every spontaneous process.

As to the choice of the term "entropy," Clausius explained:

*"I propose, accordingly, to call S the **entropy** of a body, after the Greek word '**transformation.**' I have designedly coined the word entropy to be similar to **energy**, for these two quantities are so analogous in their physical significance, that an analogy of denominations seems to me helpful."*

The choice of this term was not entirely appropriate.[13] However, during the time it was chosen, the meaning of entropy was not clear. It was a well-defined quantity, and one could calculate changes in entropy for many processes without giving a second thought to the *meaning* of entropy. Perhaps, there is no "deeper" meaning to entropy, perhaps entropy is just another physical quantity, such as volume and energy which do not have any "deeper" meaning. In fact, there are many scientists who use the concept of entropy successfully and who do not care for its meaning, or even if it has a meaning at all.

At this stage we might be content to have a quantity that is well-defined in thermodynamics. The term itself might not be appropriate, but it has been with us for over a hundred years, and whatever its meaning is, in either ancient or modern Greek, it will probably stay with us. However, once we do this we should be careful not to use the same term for another concept as this will cause huge confusion which is what happened exactly when Shannon himself acceded to the suggestion to rename his measure entropy.

Notwithstanding the enormous success and the generality of the Second Law, Clausius made one further generalization of the Second Law.

The entropy of the universe always increases

This formulation is an unwarranted *over-generalization*. Neither Clausius' definition, nor any other definition of entropy is applicable to the entire universe. It is unfortunate that the meaningless concept of the "entropy of the universe" appears quite often in recent popular science books.

As we noted earlier, Clausius did not define the entropy function, nor did he provide a method of calculating the *value* of the entropy for any system at equilibrium. This could be achieved by using the third law of thermodynamics. The application of the third law in calculating the absolute values of the entropy from experimental data (on heat capacity and heat of phase transitions) will not be discussed here. This procedure involves integration of experimental data from temperatures as close as possible to absolute zero, to the final state of interest. This is referred to as the experimental absolute value of the entropy.

On the other hand, if we assume that the Boltzmann definition is valid for all temperatures, including absolute zero $(T = 0K)$, then Boltzmann's equation implies that at absolute zero temperature, the number of states of the system is one, i.e. the entropy of the system is zero.

Indeed, for many systems for which one calculates the entropy, both from Boltzmann's equation, and from experimental data, one obtains good agreement between the two. However, there are many exceptions. Some examples are discussed in Ben-Naim (2017c).

To summarize, Clausius' definition in Eq. 2.1 requires the system to be at thermal equilibrium. However, in order to calculate finite changes in entropy or absolute entropies based on the Second Law, one must apply the concept of entropy to systems at equilibrium. It is clear that since entropy is defined for well-defined systems at equilibrium, it cannot be a function of time. In fact, time does not feature in any calculation of thermodynamic quantities in general, or entropy changes in particular.

2.2   The Boltzmann definition

Towards the end of the 19th century, the atomistic theory of matter was firmly consolidated. Back then, a majority of scientists believed that matter consists of small units called atoms and molecules. A few persistently rejected that idea arguing that there was no proof of the existence of atoms and molecules; no one has seen any atom or a molecule! Therefore, they justifiably claimed that the existence of atoms and molecules was mere speculation.

On the other hand, the so-called *kinetic theory of heat*, which was based on the assumption of the existence of atoms and molecules, had scored a few impressive gains. First, the pressure of a gas was successfully explained as arising from the molecules bombarding the walls of the container. Then came the interpretation of temperature in terms of the kinetic energy of the molecules which was a remarkable achievement that supported and lent additional evidence for the atomic constituency of matter, but fell short of a proof.

Remember that both pressure and temperature are measurable quantities. We can feel both of them on our finger tips, but neither the measurements, nor the sensation we feel on our finger tips give us any hint that these quantities are manifestations of the motions of a huge number of tiny particles.

Furthermore, the concept of *heat*, which was believed to be a kind of fluid that flows from hot to a cold body, was also interpreted in terms of the energies of all the individual molecules. Under this interpretation the First Law of Thermodynamics is nothing but an extension of the principle of conservation of energy, which now also embraces heat, or thermal energy as just another form of energy.

Thus, while the kinetic theory of heat was successful in explaining the concepts of pressure, temperature and heat, it was left lagging behind entropy and the Second Law of Thermodynamics.

Boltzmann picked up the challenge and defined the entropy in terms of the *total of number microstates* of a system consisting of a huge number of particles, but characterized by the macroscopic parameters of energy E, volume V and number of particles N.

What are these "number of microstates," and how are they related to entropy?

Consider a gas consisting of N simple particles in a volume V, each particle's microstate may be described by its location vector $\boldsymbol{R}_i$ and its velocity vector $\boldsymbol{v}_i$. By simple particles we mean particles having no internal degrees of freedom. Atoms such as argon, neon and the like are considered as simple. They all have internal degrees of freedom, but these are assumed to be unchanged in all the processes we discuss here. Assuming that the gas is very dilute so that interactions between the particles can be neglected, then, all the energy of the system is simply the sum of the kinetic energies of all the particles.

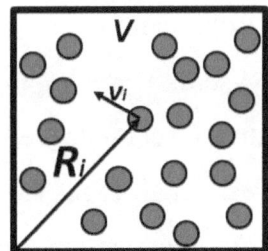

Figure 2.3. Twenty particles in a box of volume V.
Each particle i, has a locational and a velocity vector.

Imagine that you have *microscopic eyes*, and you could see the particles rushing incessantly, colliding with each other, and with the walls from time to time. Clearly, there are infinite *configurations* or *arrangements*, or *microstates* of the particles which are consistent with the requirements that the total energy is a constant, and that they are all contained within the box of volume V. One such configuration is shown in Figure 2.3. Each particle is specified by its location $\boldsymbol{R}_i$ and its velocity $\boldsymbol{v}_i$. Thus, a description of all the locations and all the velocities of the particles consist of a *microstate* of the system. In quantum mechanics one usually defines W, as the total number of quantum mechanical solutions of the Schrödinger equation for a system described by (E, V, N).

Without getting bogged down with the question of how to estimate the total number of arrangements, it is clear that this is a huge number, far "huger" than the number of particles which is of the order $N \approx 10^{23}$. Boltzmann postulated the relationship which is now known as the Boltzmann entropy

$$S = k_B \log W \tag{2.2}$$

where k_B is a constant, now known as the Boltzmann constant, and W is the number of accessible microstates of the system. Here, log is the natural logarithm. At first glance, Boltzmann's entropy seems to be completely different from Clausius' entropy. Nevertheless, in all cases for which one can calculate changes of entropy one obtains agreements between the values calculated by the two methods.

Boltzmann's entropy was a bitter pill to swallow, not only by those who did not accept the atomic theory of matter, but also by those who accepted it. The criticism was not focused so much on the definition of entropy, but rather on the formulation of the Second Law of Thermodynamics. Boltzmann explained the Second Law as a probabilistic law. In Boltzmann's words:

"… a system…when left to itself, it rapidly proceeds to disordered, most probable state."

"Most probable state." This statement was initially shocking to many physicists. Probability was totally foreign to physical reasoning. Physics was built on the foundation of deterministic and absolute laws, no provisions for exceptions. The macroscopic formulation of the Second Law was absolute - no one has ever observed a single violation of the Second Law. Boltzmann, on the other hand, insisted that the Second Law is only *statistical*, entropy increases *most* of the time, not *all* the time. The decrease in entropy is not an *impossibility* but is only highly improbable.[14] We shall further criticize this view of entropy in the following chapters of this book.

At the time when Boltzmann proclaimed the probabilistic approach to the Second Law, it seemed as if this law was somewhat *weaker* than the other laws of physics. All physical laws are absolute and no exceptions are allowed. The Second Law, as formulated by Clausius was also absolute. On the other hand, Boltzmann's formulation was not absolute – exceptions were allowed. Much later came the realization that the admitted non-absoluteness of Boltzmann's formulations of the Second Law, was in fact more *absolute* than the absoluteness of the macroscopic formulation of the Second Law, as well as of any other law of physics for that matter. See Ben-Naim (2007, 2012). It should be noted that although Boltzmann was right in claiming that the system tends to the most probable state, he erred when he claimed that the system tends to a disordered state. More details on this in Ben-Naim (2012).

There is another quantity which is sometimes also referred to as Boltzmann's entropy. This is the H-function. We shall discuss this function in Chapter 6.

Boltzmann's entropy, as defined in equation (2.2), has raised considerable confusion regarding the question of whether entropy is, or isn't a subjective quantity.

One example of this confusion which features in many popular science books is the following: Entropy is assumed to be related to our knowledge of the state of the system. If we know that the system is at some specific state, then the entropy is zero. Thus, it seems that the entropy is dependent on whether one knows or does not know in which state (or states) the system is.

This confusion arose from misunderstanding W which is the total number of accessible microstates of the system. If $W = 1$, then the entropy of the system is indeed zero (as it is for many substances at 0K). However, if there are W states and we know in which state the system is, the entropy is still $k\ln W$. This kind of argument led some authors to reject the "informational interpretation" of entropy. For details and examples see Ben-Naim (2017a, 2017c)

Another kind of confusion, which is common in popular science books written by authors who accept the informational interpretation of entropy.

Since the Shannon measure of information (SMI) is confused with "information," and since information may be subjective, one concludes that also entropy must be a subjective quantity. As we shall see in the next chapters, entropy is related to the probability distribution of locations and momenta at equilibrium. This distribution (at equilibrium) is *fixed* by the macroscopic state of the system. For example, for an ideal gas of N atoms at temperature T, and in a volume V, the distribution of locations and momenta is determined. So is the entropy of the system, which we can write as $S = S(T, V, N)$. This entropy is independent of *our knowledge* of the distribution of locations and momenta.

Although it was not explicitly stated in the definition, the Boltzmann entropy applies to an equilibrium state. In statistical mechanics this entropy applies to the so-called micro-canonical ensemble, i.e. to systems having a fixed energy E, volume V, and number of particles N (N could be a vector comprising the number of particles of each species ($N_1, \dots, N_c$). It is also clear that the entropy of an ideal gas applies to a well-defined system at equilibrium.

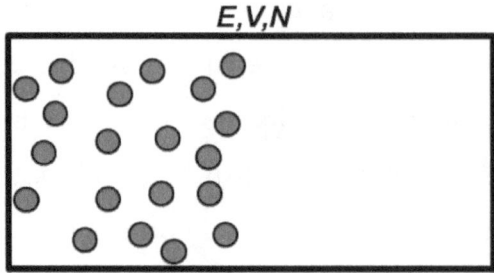

Figure 2.4. *N* particles in a box of volume *V,* and energy *E.*
This particular configuration, obtained right after the
removal of the partition, is one of the total number of
possible configurations of the system.

When one refers to W, as the number of accessible microstates, these also include microstates which we may refer to as non-equilibrium states. For instance, in a state of N particles confined to a volume V, and having total energy E, the state shown in Figure 2.4 is reckoned as one of *all* the accessible states of this system. These states might be viewed as non-equilibrium states in the sense that if we measure the temperature, pressure, or chemical potential at different places in the system we shall find out that the parameters change with time. However, these states may also be seen as arising from fluctuations about the equilibrium state. Such fluctuations are so rare that we do not observe them in practice. When we define the entropy of such a system it includes *all possible microstates.* However, in practice only states which are very close to the equilibrium states are observable. The applicability of entropy to equilibrium states will be clearer when we derive the same entropy function of an ideal gas from the SMI.

2.3 The ABN definition based on Shannon's measure of information

This is a relatively recent definition of entropy.[15] It is however, superior to both the Clausius and the Boltzmann definitions. Unlike the Clausius definition, which provides only a definition of changes in entropy, the present one provides the *entropy function* itself. Unlike Boltzmann's definition, which is strictly valid for isolated systems, and does not provide a simple intuitive interpretation, the present one is more general and provides a clear, simple and intuitive interpretation of entropy. It is more general in the sense that it relates the entropy to probability distributions, rather than to the number of microstates. One final "bonus" is afforded by this definition of entropy. It not only removes any trace of mystery associated with entropy, but it also expunges the so-called irreversibility paradox.

Since most scientists are unfamiliar with the SMI we shall devote part of Chapter 3 to introducing the SMI, its interpretations, and its properties. It is my conviction that any educated person should be exposed to the SMI, not because it is related to entropy, but because of its extreme generality, usefulness, and beauty.

The SMI is a very useful concept. It was found useful in many branches of science. Here, we present only the qualitative aspects of the concept of the Shannon measure of information, present a few simple examples, then describe the four steps leading from the SMI to the thermodynamic entropy. We shall see that the entropy is, up to a multiplicative constant, nothing but a particular case of SMI. For more details, see Ben-Naim (2017c).

Chapter 3. Derivation of the Entropy Function from Shannon's Measure of Information (SMI)

In this chapter we derive the entropy of an ideal gas of simple particles from Shannon's Measure of Information. To do this we shall start with a brief introduction to SMI, its properties, and its interpretations. Next, we use the maximum SMI principle to calculate the entropy of an ideal gas. This procedure provides a simple reliable and proved interpretation of entropy. We also note that this interpretation can also be extended to system of interacting particles.

Once we have the *entropy function* we can use this function as a *new definition* of entropy, a definition which is more general than either Clausius' or Boltzmann's. Yet, results calculated from the new definition conform with the results calculated with the older definitions.

3.1 Shannon's Measure of Information (SMI)

Although we shall not discuss it here, the reader should be aware of the fact that there is an immense difference between the *concept* of *information* and the *measure* of *information*. We also note that Shannon erred when he named his measure *entropy*. This has caused great confusion in both information theory and in thermodynamics.

In 1948, Shannon published a landmark article titled, "A Mathematical Theory of Communication."

A year later a slightly expanded work was published as a book by Shannon and Weaver (1949) entitled, "The Mathematical Theory of Communication."

Note the minor, yet significant difference in the two titles. More importantly, one should note that the article, as well as the book presents a *"theory of communication,"* not a *"theory of information."* Notwithstanding the difference between "communication" and "information," Shannon's work is now considered to be the cornerstone of what is now referred to as "Information Theory" (IT).   Shannon was interested in a theory of *communication of information*, not information itself. This is very clear to anyone who read through Shannon's original article. In fact, in the introduction to the book, we find:

"The word information, in this theory, is used in a special sense that must not be confused with its ordinary usage. In particular, information must not be confused with meaning.

In fact, two messages, one of which is heavily loaded with meaning and the other of which is pure nonsense, can be exactly equivalent, from the present viewpoint, as regards information.

The word communication will be used here in a very broad sense to include all of the procedures by which one mind may affect another. This, of course, involves not only written and oral speech, but also music, the pictorial arts, the theatre, the ballet, and in fact all human behavior.

The fundamental problem of communication is that of reproducing at one point either exactly or approximately a message selected at another point. Frequently the messages have meaning…These semantic aspects of communication are irrelevant to the engineering problem."

It is clear that what is being referred to as IT is not a *theory of information*! In fact, there is no theory of information which takes into account the *meaning*, the *value*, the *significance*, etc. of the information.

The best way to appreciate what IT is all about is to read Shannon's own words. In section 6 of the article titled: "Choice, Uncertainty and Entropy," we find:

Suppose we have a set of possible events whose probabilities of occurrence are $p_1, p_2, \cdots, p_n$. These probabilities are known but that is all we know concerning which event will occur. Can we find a measure of how much "choice" is involved in the selection of the event or how uncertain we are of the outcome?

If there is such a measure, say, $H(p_1, p_2, \ldots, p_n)$, it is reasonable to require of it the following properties:

1. H should be continuous in the p_i.

2. If all the p_i are equal, $p_i = \frac{1}{n}$, then H should be a monotonic increasing function of n. With equally likely events there is more choice, or uncertainty, when there are more possible events.

3. If a choice be broken down into two successive choices, the original H should be the weighted sum of the individual values of H.

Then Shannon proved the theorem:

The only H satisfying the three assumptions above has the form:

$$H = -K \sum p_i \log p_i \qquad (3.1)$$

Let us quote another paragraph from Shannon's article:

This theorem, and the assumptions required for its proof, are in no way necessary for the present theory. It is given chiefly to lend a certain plausibility to some of our later definitions. The real justification of these definitions however, will reside in their implications.

Quantities of the form $H = -K \sum p_i \log p_i$ (the constant K merely amounts to a choice of a unit of measure) play a central role in information theory as measures of information, choice and uncertainty. The form of H will be recognized as that of entropy as defined in certain formulations of statistical mechanics where p_i is the probability of a system, being in cell i of its phase space. H is then, for example, the H in Boltzmann's famous H theorem. We shall call $H = -K \sum p_i \log p_i$ the entropy of the set of probabilities $p_i, \dots, p_n$.

First, note that Shannon describes H as a "measure of information, choice and uncertainty." All these are valid interpretations of the quantity H, as defined above. We shall devote more time to these interpretations in the following sections of this chapter.

Second, note that Shannon did not explicitly search for a measure of information. Instead, he formulated his problem in terms of a *probability distribution $p_1, \dots, p_n$*. He sought a measure of how much "choice," or "uncertainty" there is in the outcome, and he later referred to the quantity H as a measure of "information, choice and uncertainty."

Shannon did not seek a measure of the general concept of information, only a *measure* of *information contained in*, or *associated with* a probability distribution. This is an important point that one should remember whenever using the term "information" either as a measurable quantity, or in connection with the Second Law of Thermodynamics.

Third, Shannon proposed three plausible properties of such a measure, *presuming* that such a measure exists. We shall discuss these properties and their plausibility in the following sections of this chapter.

Finally, note carefully that Shannon was not interested in thermodynamics in general, or in *entropy* in particular. However, he noted that "*the form of H will be recognized as that of entropy as defined in certain formulations of statistical mechanics…*" Therefore, he suggested calling H "the entropy of the set of probabilities $p_1, … , p_n$."

Indeed, the *form* of the function H is the same as the *form* of the entropy as used in statistical mechanics. However, the fact that the *form* of H is the same as the form of the entropy in statistical mechanics *does not* imply that H is entropy. Also, it is not true that H, in the Boltzmann H-Theorem is entropy. We will further discuss this point in Chapter 6. For the moment we will study the SMI without any reference to entropy. However, the reader should be aware of the fact that in many applications of the concept of SMI, the concept of entropy has also been involved. This fact has caused great confusion in both information theory and thermodynamics.

SMI is a very general concept. It is defined on *any discrete distribution function*. Examples are the outcomes of throwing a dice and the frequencies of the appearance of letters in the alphabet in certain languages. There is a vast range of fields in which the quantity H is definable. This has made SMI a very useful tool in many fields of research.

The entropy, as we shall soon see is defined only on a narrowly specified set of probability distributions. When H is applied to those distributions used in statistical mechanics, it is identical, up to a multiplicative constant, with the statistical mechanical entropy. Thus, the statistical mechanical entropy is a particular case of SMI, but the SMI is in general not the entropy. Unfortunately, confusion of the two concepts abounds. The source of this confusion is probably due to von Neumann's suggestion to Shannon to name the quantity H "entropy." The story is told by Tribus and McIrvine (1971):

What's in a name? In the case of Shannon's measure the naming was not accidental. In 1961 one of us (Tribus) asked Shannon what he had thought about when he had finally confirmed his famous measure. Shannon replied: "My greatest concern was what to call it. I thought of calling it 'information,' but the word was overly used, so I decided to call it 'uncertainty.' When I discussed it with John von Neumann, he had a better idea. Von Neumann, told me, 'You should call it entropy, for two reasons. In the first place your uncertainty function has been used in statistical mechanics under that name, so it already has a name. In the second place, and more important, no one knows what entropy really is, so in a debate you will always have the advantage.

In this book, we will refer to the quantity H defined above as the Shannon's measure of information (SMI).

There are essentially two kinds of applications of the SMI. The first; when we know the probability distribution of the outcomes of an experiment, we can calculate the SMI and use it to interpret other quantities. Perhaps, the most important application of this type is the interpretation of the longstanding mysterious and resistant-to-understanding quantity called *entropy*. We shall discuss the power of the SMI in demystifying entropy and the Second Law in the following sections.

The second application is the so-called MaxEnt method, which we shall refer to as the MaxSMI method. Here, we have an experiment (or a game, or a random variable), but we do not know the probability distribution of the outcomes. The question is how to find out the "best" distribution which is consistent with whatever knowledge we have about the experiment. We shall also use the MaxSMI method to derive the entropy function of an ideal gas.

In this chapter we define the SMI for any given distribution $p_1, \ldots, p_n$ by:

$$H = -\sum_{i=1}^{n} p_i \log p_i \tag{3.2}$$

The logarithm is to the base 2.

3.2 Various Interpretations of SMI

In this section we discuss three interpretations of the SMI. The first is an *average* of the uncertainty about the outcome of an experiment; the second, a measure of the average *unlikelihood*; and the third, a measure of information. It is ironic that the "informational" interpretation of the SMI is the most difficult to see, and as a result it is also the one most commonly misused. We shall use the letter H for the quantity defined above and refer to it simply as SMI. Note that the SMI in equation (3.1) has the form of an *average* quantity. However, this is a very special average. It is an average of the quantity $-\log p_i$ using the probability distribution $p_1, \ldots, p_n$.

3.2.1 The uncertainty meaning of SMI

The interpretation of H as an *average uncertainty* is quite popular. This interpretation is derived directly from the meaning of the probability distribution.

Suppose that we have an experiment yielding n possible outcomes with probability distribution $p_1, \ldots, p_n$. If, say, $p_i = 1$, then we are *certain* that the outcome i occurred or will occur. For any other value of p_i (greater than zero but less than one), we are *less certain* about the occurrence of the event i. *Less certainty* can be translated to *more uncertainty*. Therefore, the larger the value of $-\log p_i$, the larger the extent of uncertainty about the occurrence of the event i. Multiplying $-\log p_i$ by p_i, and summing over all i, we get an *average uncertainty* about *all* the possible outcomes of the experiment.

We should add here that when $p_i = 0$, we are *certain* that the event i *will not* occur. It would be awkward to say in this case that the *uncertainty* in the occurrence of i is zero. Fortunately, this awkwardness does not affect the value of H. Once we form the product $p_i \log p_i$, we get zero when $p_i = 0$.

Finally, we note that whenever we say that the SMI is a measure of uncertainty, we mean uncertainty with respect to all the outcomes of an experiment, in the sense discussed above. For instance, when we throw a die, we can talk of many uncertainties; the color, mass, form, etc. of the die. These uncertainties are not the SMI of the experiment of throwing a die. Unfortunately, you can find in many popular science books a description of the SMI (as well as entropy) as uncertainty, without specifying what that uncertainty refers to.

3.2.2 The unlikelihood interpretation

A slightly different but still useful interpretation of H is in terms of *likelihood* or *expectedness*. These two are also derived from the meaning of probability. When p_i is small, the event i is unlikely to occur, or its occurrence is less expected. When p_i approaches 1, the occurrence of i becomes more likely or more expected. Since $\log p_i$ is a monotonically increasing function of p_i, we can say that the larger the value of $\log p_i$, the larger the likelihood or the larger the expectedness for the event. Since $0 \leq p_i \leq 1$, we have $-\infty \leq \log p_i \leq 0$. The quantity $-\log p_i$ is thus a measure of the *unlikelihood* or the *unexpectedness* of the event i. Therefore, the quantity $H = -\sum p_i \log p_i$ is a measure of the *average unlikelihood,* or *unexpectedness* of the entire set of the outcomes of the experiment.

3.2.3 The meaning of the SMI as a measure of information

As we have seen, both the uncertainty and the unlikelihood interpretation of H are derived from the meaning of the probabilities p_i. The interpretation of H as a measure of information is somewhat trickier and less straightforward. It is also more interesting since it conveys a different kind of *information* on the Shannon measure of *information.* As we already emphasized, the SMI is not *information.* Also, it is not a measure of any piece of information, but of a very particular kind of information. The confusion of SMI with information is almost the rule, not the exception by scientists and non-scientists alike.

Some authors assign to the quantity $-\log p_i$ the meaning of information (or self-information) associated with the event i. The idea is that if an event is rare, i.e. p_i is small and hence $-\log p_i$ is large, one obtains "more information" when one knows that the event has occurred.

Both p_i and $\log p_i$ are measures of the uncertainty about the occurrence of an event. They do not measure *information* about the events. Therefore, we do not recommend referring to $-\log p_i$ as "information" (or self-information) associated with the event i. Hence, H should not be interpreted as *average information* associated with the experiment. Instead, we assign the "informational" definition directly to the quantity H, rather than to the individual events.

It is sometimes said that removing the *uncertainty* is tantamount to obtaining *information*. This is true for the entire experiment, i.e. to the entire probability distribution, not to individual events.

Suppose that we have an unfair die with probabilities $p_1 = \frac{1}{10}, p_2 = \frac{1}{10}, p_3 = \frac{1}{10}, p_4 = \frac{1}{10}, p_5 = \frac{1}{10}$ and $p_6 = \frac{1}{2}$. Clearly, the uncertainty we have regarding the outcome $i = 6$ is less than the uncertainty we have regarding any outcome $i \neq 6$. When we carry out the experiment and find the result, say $i = 3$, we remove the uncertainty we had about the outcome before doing the experiment. However, it would be wrong to argue that the *amount* of information we obtained is larger or smaller if another outcome had occurred. Note also that we discuss here the *amount* of information, and not *the information* itself. If the outcome is $i = 3$, the information we obtained is: The outcome is "3." If the outcome is $i = 6$, the information is: The outcome is "6." These are different pieces of information, but one cannot claim that one is larger or smaller than the other.

We emphasize again that the interpretation of H as an *average uncertainty* or *average unlikelihood* is derived from the meaning of each term $-\log p_i$. The interpretation of H as a measure of information is not associated with the meaning of each probability p_i, but with the *entire distribution* $p_1, \ldots, p_n$.

In this section, we describe in a qualitative way the meaning of H as a *measure* of *information associated with the entire experiment.* We shall say more about this interpretation of H in the following sections.

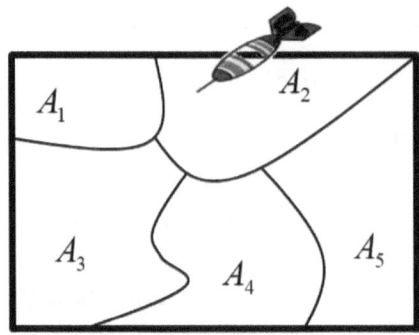

Figure 3.1. A 20Q-game with five events.
The event *i* is: "The dart is in region A_i
The probability of the event *i* is A_i / A

Consider any experiment or a game having n outcomes with probabilities $p_1, ..., p_n$. For concreteness, suppose we throw a dart at a board, Figure 3.1. The board is divided into n regions, of areas $A_1, ..., A_n$. We know that the dart hit one of these regions. We assume that the probability of hitting the ith region is $p_i = A_i/A$, where A is the total area of the board.

Now that the experiment had been carried out, you have to find out where the dart hit the board. You know that the dart hit the board, and you know the probability distribution $p_1, ..., p_n$. Your task is to find out in which region the dart is by asking binary questions, i.e. questions answerable by Yes, or No.

Clearly, since you do not *know* where the dart is you *lack information* on the location of the dart. In order to acquire this information, you have to ask questions. We are interested in the *amount* of *information* contained in this experiment. One way of measuring this amount of information is by the *number* of *questions* you need to ask in order to obtain the required information.

As anyone who has played the 20-question (20Q) game knows, the number of questions you need to ask depends on the *strategy* for asking questions. At this point we shall not elaborate on the question of strategies, see Ben-Naim (2006b, 2017c). For now, we are only interested in a measure of the "amount of information" contained in this experiment. It turns out that the quantity H, to which we referred to as Shannon's Measure of Information (SMI) provides us with a measure of this information in terms of the minimum number of questions one needs to ask in order to find the location of the dart, given the *probability distribution* of the various outcomes.

For a general experiment with n possible outcomes, having probabilities $p_1, ..., p_n$, the quantity H will be shown to be a measure of how "difficult" it is to find out which outcome has occurred given that an experiment was carried out. As it is easy to prove, for experiments having the same total number of outcomes n, but with different probability distributions, the amount of information (measured in terms of the number of questions) is different. In other words, knowing the probability distribution gives us a "hint" or some partial information on the outcomes. This is the reason why we refer to H as a measure of the amount of information *contained* in, or *associated* with, a given probability distribution. We emphasize again that the SMI is a measure of information associated with the *entire* distribution, and not with individual probabilities.

3.3 A simple 20Q game

In its most general form the 20Q game may be described as follows:

We start with n objects (or persons, events, or whatever it may be). I choose one object, and you have to find out which object I chose by asking binary questions, i.e. questions that are answerable by Yes or No.

Figure 3.2 A simple 20Q-game with 32 objects

Figure 3.2 shows an example of a 20Q game. We have n objects $A_1, A_2, ..., A_n$ and the corresponding probabilities $(p_1, ..., p_n)$. Before one asks the first question, one must divide the total of n objects into two groups A and B, and ask: Is the object in group A (or in group B)? If the answer is Yes, one proceeds to divide the objects in A, into two groups, and so on. This procedure is shown schematically in Figure 3.3.

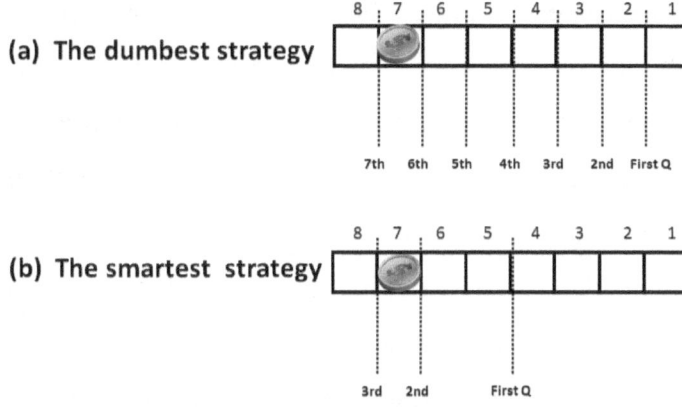

Figure 3.3. Two strategies of asking questions:
(a) The dumbest, and (b) The smartest.

Before we discuss specific games, two comments are in order.

First, in a normal 20Q game we do not explicitly specify all the objects in the set from which the object is chosen. For instance, I can choose a person from a group attending the party, or a person from a specific city, or country, or even from whole world. Clearly, if I choose a person or an object you have never heard of, you will not be able to guess who the person is, or what the object is.

Therefore, in a "fair" game we must agree on the group of objects or persons from which I shall choose one. Clearly, this group must be finite, otherwise there is a chance that you will not be able to guess what the object is in a finite number of questions – and the party can go on and on.

Second, in an actual parlor game we never specify the probabilities of the events. More precisely we do not specify with what probability I chose a specific person or object. It is usually assumed implicitly that the objects are chosen with equal probabilities, but in fact that is never the case. For instance, if I had to choose a person, I might be biased and choose a person that I am familiar with, or like, or hate, or someone I know whose identity might be difficult to guess, etc. On the other hand, if you know me, you might use this knowledge in order to venture a better guess about the person I am most likely to choose. Thus, in such a game there are all kinds of psychological elements that can play out in the process of *choosing* the person, as well as in guessing the chosen person. Therefore, in the following we shall make the 20Q game more precise, objective and hence, more susceptible to a mathematical study.

As anyone who has played the 20Q game knows, there are various strategies of asking questions. It is also intuitively clear that some strategies are "better" than the others, in the sense that they lead to the required information with fewer questions. We shall briefly examine these strategies and their differences with a few examples. It is interesting to note that young children learn at an early age the best strategy of asking questions. For details, see Ben-Naim (2010).

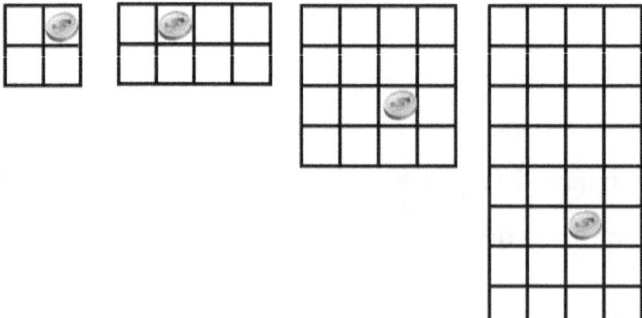

Figure 3.4. A coin hidden in one of n boxes, with n = 4, 8, 16, 32

In Figure 3.4 we show a few games with different numbers of outcomes; 4, 8, 16, 32. (The game could be a coin hidden in one of n boxes, or a board divided into n regions on one of which the dart hit). It is clear that asking specific questions (such as: "Is the coin in box k?") is not the most efficient way of getting the required information (on where the coin is). One can prove that the most efficient (or the "smartest") strategy of asking questions is to divide the entire set of outcomes into two groups having equal probabilities. In such a case one gets one *bit* of information for any binary question. As we shall soon see, the *bit* is the maximum amount of information one can get by asking a binary question, using this strategy will require a minimum number of questions in order to find out where the coin is (or where the dart hit). Note that it is not always possible to divide the entire set of outcomes into two equally probable groups. However, one can show that by using the smartest strategy one needs to ask approximately $\log_2 n$ questions to find out the one outcome out of n possible outcomes. Figure 3.5 shows two games, one with uniform and one with non-uniform distribution of outcomes. One can prove that the uniform game is the most difficult game to play (in the sense that one will need, on average, to ask more questions). In any game, uniform or non-uniform, the SMI gives the average number of questions one needs to ask in order to find out one out of n possible outcomes. This is the basis for the interpretation of SMI as a measure of information. For details, see Ben-Naim (2008, 2017b).

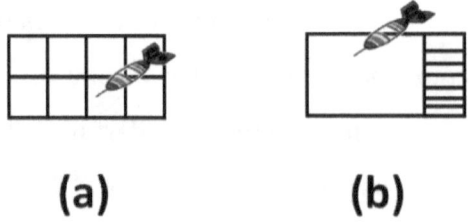

(a)   **(b)**

Figure 3.5. Two games with (a) Uniform, and (b) Non-uniform distribution

3.4 Properties of Shannon's Measure of Information

In this section we discuss some of the main properties of the SMI. Some of these properties were *assumed* by Shannon when he sought a measure of information, or of uncertainty. In particular, we shall see how the SMI is related to the number of questions one needs to ask in the familiar 20Q game. We shall also introduce the continuous analogue of the SMI and derive from it three important probability densities.

We start with the general definition of SMI for a finite set of events:

As we discussed in section 3.1, Shannon sought a *measure* of *information* associated with a probability distribution. The probability distribution is associated with set of events $A_1, A_2, \ldots, A_n$ which is *complete* and the events are pairwise *mutually exclusive*. This means that one, and only one of the events $A_1, \ldots, A_n$, has occurred, or will occur. The corresponding probabilities are: $p_1, \ldots, p_n$ with $p_i \geq 0$ and $\sum_{i=1}^{n} p_i = 1$. Sometimes, we shall refer to the distribution associated with an experiment, meaning the probabilities of all the outcomes of the experiments.

Thus, for any probability distribution, the SMI is defined by

$$H(p_1, \ldots, p_n) = -\sum_{i=1}^{n} p_i \log p_i \qquad (3.3)$$

The base of the logarithm is usually chosen to be 2. However, in some applications we might use any other base we wish. Note that the quantity H has the *structure* of an average quantity. However, this is a very special kind of average. In general, an average quantity is defined by:

$$\langle M \rangle = \sum_{i=1}^{n} p_i M_i \qquad (3.4)$$

Here, M_i is some quantity associated with the event A_i and p_i is the probability of occurrence of this event. The quantity $\langle M \rangle$ is referred to as the *average*, the *mean*, or the *expected* value of the quantity M.

Comparing equations (3.3) and (3.4) we see that H may be interpreted as an average of the quantities $-\log p_i$, each of which is related to the probability p_i. In this sense, H may be said to be a *purely probabilistic*.

In Equation (3.3) we used the notation $H(p_1, \ldots, p_n)$ to denote the functional dependence of H on the variables $p_1, \ldots, p_n$. Sometimes, we use the notation $H(X)$ to denote the quantity H defined for an experiment or a random variable X. In more general cases when we have two or more random variables, say X_1, X_2, we denote by $H(X_1, X_2)$ the SMI associated with the two random variables, X_1, X_2. In this notation $H(X_1, X_2)$ does not denote a functional dependence of H on the "variables" X_1, X_2. If p_{ij} is the joint probability of the event $(X_1 = x_i)$, and $(X_2 = x_j)$, then $H(X_1, X_2)$ is a shorthand notation for the function of $n \times m$ independent variables p_{ij}

$$H(p_{11}, p_{12}, \ldots, p_{nm}) = -\sum_{i=1}^{n} \sum_{j=1}^{m} p_{ij} \log p_{ij} \qquad (3.5)$$

Similar generalizations can be applied to any number of random variables. In most of the applications discussed in this book we defined the SMI for a finite number of events (or outcomes). In some applications we use an infinite number of events, or even a continuous case. See section 3.4.4.

3.4.1 The case of an experiment having two outcomes; definition of a unit of information

The simplest case for which the SMI can be defined is the case of the two outcomes, e.g. the outcomes H and T in the tossing of a coin. This case is important for several reasons:

1. It is the simplest case for which we can understand the meaning of H as a measure of uncertainty or as a measure of information.

2. It is the simplest case with the help of which we can visualize the properties of the function $H(p_1, 1 - p_1)$, or simply $H(p)$, where $p = p_1$.

3. It is the basis for *defining* the *unit* of *information* the *bit*.

From the general definition of H in Eq. (3.3) we can write for the case $n = 2$ the function

$$H(p_1, p_2) = -p_1 \log p_1 - p_2 \log p_2 \qquad (3.6)$$

Since $\sum p_i = 1$, H is a function of only one independent variable. We plot this function in Figure 3.6. For convenience, we write this function as

$$H(p) = -p \log p - (1 - p)\log(1 - p) \qquad (3.7)$$

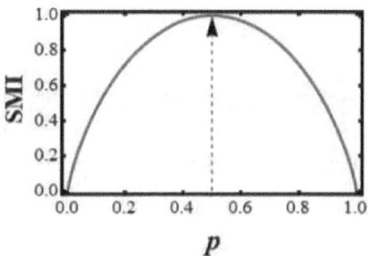

Figure 3.6. The SMI for the case of a two-outcome experiment.

This function is defined in the range $0 \leq p \leq 1$. It has a *single* maximum at $p = \frac{1}{2}$. It is concave downward, and it is zero at both end points: $p = 0$ and $p = 1$.

The continuity of the function $H(p)$ is obvious. To see that it has a single maximum, we take the two derivatives of $H(p)$ (note that $\log_2 p = \ln p / \ln 2$, where ln is the natural logarithm).

$$(\ln 2)\frac{dH}{dP} = \ln\frac{1-p}{p} \tag{3.8}$$

$$(\ln 2)\frac{d^2H}{dp^2} = \frac{-1}{p(1-p)} \tag{3.9}$$

The condition for an extremum is:

$$\frac{dH}{dP} = 0 \tag{3.10}$$

or equivalently

$$p = (1 - p) \tag{3.11}$$

Thus, $H(p)$ has a *single* solution at

$$p = \frac{1}{2} \tag{3.12}$$

Clearly, the extremum at $p = \frac{1}{2}$ is a maximum since the second derivative, Eq. (3.9) is always negative in the range $0 < p < 1$.

The value of the function of $H(p)$ at the maximum is:

$$H_{max} = H\left(p = \frac{1}{2}\right) = 1 \tag{3.13}$$

The value of $H(p)$ at each of the extreme points $p = 0$ and $p = 1$ is zero. By using L'Hopital's theorem it is easy to show that the product $p\log p$ (in any base) tends to zero, when p tends to zero

$$\lim_{p \to 0}(p\ln p) = \lim_{p \to 0}\frac{\ln P}{1/p} = \lim_{p \to 0}\frac{\frac{d}{dp}[\ln p]}{\frac{d}{dp}\left[\frac{1}{p}\right]} = \lim_{p \to 0}\frac{1/p}{-1/p^2} = 0 \tag{3.14}$$

Obviously, the curve in Figure 3.6 is everywhere concave downwards (i.e. the second derivative is negative) in the entire range $0 \le p \le 1$.

We define the *unit* of *information*, the *bit*, as the *amount of information* you obtain when you ask a binary question regarding two outcomes with *equal probabilities*, i.e. this is the value of $H(p)$ at $p = \frac{1}{2}$, $H\left(\frac{1}{2}\right) = -\frac{1}{2}\log\frac{1}{2} - \frac{1}{2}\log\frac{1}{2} = 1$. This is also the maximum information you can obtain for a binary question, see also Figure 3.6. (Note that in communication theory one uses another definition of the bit. For details, see Ben-Naim (2017c).

3.4.2 The Shannon measure of information for a finite number of outcomes

Originally, Shannon assumed that a measure of uncertainty or a measure of information must have certain properties. Through the years, various modifications of these properties have been suggested. One convenient list of properties (sometimes referred to as axioms) is the following:

1. The function $H(p_1, ..., p_n)$ is a continuous function of all the variables p_i, where $0 \leq p_i \leq 1$ and $\sum_{i=1}^{n} p_i = 1$.

2. When $p_i = p = \frac{1}{n}$ for all $i = 1, 2, ..., n$, the function $H(p_1, ..., p_n) = f(n)$ is a monotonically increasing function of n.

3. The function $f(n)$ has the property

$$f(n \times m) = f(n) + f(m) \tag{3.15}$$

4. The function $H(p_1, ..., p_n)$ fulfills the equality

$$H(p_1, ..., p_n) = H(p_A, p_B) + p_A H\left(\frac{p_1}{p_A}, ..., \frac{p_r}{p_A}\right) p_B H\left(\frac{p_{r+1}}{p_B}, ..., \frac{p_n}{p_B}\right) \tag{3.16}$$

where we denoted by $p_A = \sum_{i=1}^{r} p_i$ and $p_B = \sum_{i=r+1}^{n} p_i$. This property is referred to as the grouping, or the consistency property.

The continuous property of the function $H(p_1, ..., p_n)$ is obvious. Also, for equally probable probabilities $p_i = p = 1/n$, the function

$$f(n) = -\sum_{i=1}^{n} p_i \log p_i = \log n \tag{3.17}$$

which is monotonically increasing function of n.

The additivity property:

$$f(n \times m) = f(n) + f(m) \tag{3.18}$$

follows from the basic property of the logarithm function, i.e.

$$\log(n \times m) = \log n + \log m \tag{3.19}$$

The last condition (4) is less intuitive. However, this property is "natural" if we want to interpret H as a measure of information. It is also important in coding theory, but the significance of this property to coding will not be discussed here.

3.4.3 The maximum of the SMI over all possible discrete distributions

We now generalize what we have found in section 3.4.1. We are given a game (or a random variable, or an experiment) having n outcomes with a distribution $p_1, \ldots, p_n$. The function $H(p_1, \ldots, p_n)$ has a *single maximum* over all possible distributions with the same n.

The mathematical problem is to find the maximum of the function

$$H(p_1, \ldots, p_n) = -\sum_{i=1}^{n} p_i \log p_i \tag{3.20}$$

subject to the constraint (or the closure condition)

$$\sum_{i=1}^{n} p_i = 1 \tag{3.21}$$

In this and in the next section, we use the natural logarithm for convenience. This choice simplifies the mathematics but does not affect the final result.

We define the auxiliary function:

$$F = H(p_1, \ldots, p_n) + \lambda\left(\sum_{j=1}^{n} p_j - 1\right) \tag{3.22}$$

where λ is a constant which will be determined later. The condition for an extremum of F is that all the partial derivatives of F with respect to each of the p_i be zeros, i.e.

$$\left(\frac{\partial F}{\partial p_i}\right)_{p_i'} = -\log p_i - 1 + \lambda = 0 \tag{3.23}$$

The symbol p_i' stands for the vector $(p_1, p_2, \ldots, p_{i-1}, p_{i+1}, \ldots, p_n)$, i.e. all the components except p_i. From eq. (3.23) we get the distribution which maximizes H;

$$p_i^* = \exp(\lambda - 1) \tag{3.24}$$

Substituting (3.24) into (3.21), we obtain

$$1 = \sum_{i=1}^{n} p_i^* = \exp(\lambda - 1) \sum_{i=1}^{n} 1 = n \exp(\lambda - 1) \tag{3.25}$$

From this equation we get λ, which is used in (3.24) to obtain

$$p_i^* = \frac{1}{n} \tag{3.26}$$

This is an important result. It states that the maximum value of H, subject to only the condition (3.21), is obtained when the distribution is *uniform*. This is a generalization of the result we have seen in section 3.4.1. It is easy to see that at the uniform distribution H has a *maximum*. This follows from the fact that the second derivative of H is always negative,

$$\frac{d^2 H}{d P_i^2} < 0 \quad \text{(for all } i) \tag{3.27}$$

The value of H at the maximum is:

$$H_{max} = -\sum_{i=1}^{n} p_i^* \log p_i^* = -\sum_{i=1}^{n} \frac{1}{n} \log \frac{1}{n} = \log n \qquad (3.28)$$

Clearly, when there are n equally likely events, the value of H is larger, the larger the number of possible outcomes.

3.4.4 The case of infinite number of outcomes

The definition of the SMI for the case of discrete infinite number of possibilities is straightforward. First, we recall that for a finite and uniform distribution, we had

$$H = \log n \qquad (3.29)$$

where n is the number of possibilities. Taking the limit, $n \to \infty$, we get

$$H = \lim_{n\to\infty} \log n = \infty \qquad (3.30)$$

This means that the amount of information contained in the game tends to infinity. Note however that the probabilities, $1/n$ tend to zero. The interpretation of this result is simple. If we have an infinite number of equally probable possibilities, then we will need to ask, on average an infinite number of questions.

For a non-uniform distribution, the quantity H might or might not exist, depending on whether the quantity

$$H = -\sum_{i=1}^{\infty} p_i \log p_i \qquad (3.31)$$

converges or diverges.

The case of a continuous distribution is problematic. If we start from the discrete case and proceed to the continuous limit we encounter some difficulties. For details, see Ben-Naim (2017c). Here, we shall follow Shannon's treatment for a continuous distribution for which a density function $f(x)$ exists. In analogy with the definition of the H function for discrete probability distribution, we define the quantity H for a continuous distribution. Let $f(x)$ be the density distribution, i.e. $f(x)dx$ is the probability of finding a particular value of the random variable between x and $x + dx$.

We defined the H function in the form:

$$H = -\int_{-\infty}^{\infty} f(x) \log f(x) dx \qquad (3.32)$$

Note carefully that in this definition $f(x)dx$ is a pure number. In general, $f(x)$ itself is not a pure number (e.g. if dx has units of length, then $f(x)$ has units of 1/length. Therefore, one must be careful in using this definition of H. See also Ben-Naim (2017b).

As we noted in the previous section, for mathematical convenience, we use in this section the natural logarithm in the definition of H.

3.4.5 Three extremum theorems on the SMI

In this section we discuss three important theorems proved by Shannon (1948). They are important for three reasons: First, they show how three fundamental distributions in probability theory and statistics arise; second, they shed new light on the meaning of equilibrium state in thermodynamics; and third, they are essential to the understanding of both entropy and the Second Law of thermodynamics.

The uniform distribution of locations

Consider a particle that is confined to a one-dimensional "box" of length L, we seek the maximum of H defined in (3.32), but with limits $(0, L)$, subject to the conditions that

$$\int_0^L f(x)dx = 1 \tag{3.33}$$

We apply the Lagrange method of undetermined multiplier (or the calculus of variation). We define the auxiliary functional

$$A[f(x)] = H[f(x)] + \lambda \int_0^L f(x)dx \tag{3.34}$$

Take the functional derivative with respect to the component $f(x')$, we obtain:

$$\frac{\delta A}{\delta f(x\prime)} = -1 - \log f(x') + \lambda \tag{3.35}$$

For details, see Ben-Naim (2017b). From (3.35), we find that the probability density $f^*(x)$ which maximizes H, subject to the condition (3.33) must satisfy the equality

$$-1 - \log f^*(x) + \lambda = 0 \tag{3.36}$$

Using the result (3.36) in (3.33), we obtain

$$1 = \int_0^L f^*(x)dx = e^{\lambda-1} \int_0^L dx = e^{\lambda-1}L \tag{3.37}$$

or equivalently

$$f^*(x) = \frac{1}{L} \tag{3.38}$$

This is the density which maximizes H subject to the condition (3.33). We shall refer to this distribution as the *equilibrium distribution* (see Chapter 6) and use the notation $f_{eq}(x)$ instead of $f^*(x)$.

Thus, the equilibrium density distribution is *uniform* over the entire length L. The probability of finding the particle at any interval, say between x and $x + dx$ is

$$f_{eq}(x)dx = \frac{dx}{L} \tag{3.39}$$

which is independent of x. This result is of course in accordance with our expectations. Since no point in the box is preferred, the probability of being found in an interval dx is simply proportional to the length of that interval. A more general result is when the density function $f(x)$ is defined in an interval (a, b), in this case the maximum SMI is obtained for the density function

$$f(x) = \frac{1}{b-a} \quad , \quad \text{for } a \le x \le b \tag{3.40}$$

and the corresponding value of the SMI is $H = \log(b - a)$

The SMI associated with equilibrium density (3.39) is

$$H_{max} = -\int_0^L f_{eq}(x) \log f_{eq}(x)dx = -\frac{1}{L}\log\frac{1}{L}\int_0^L dx = \log L \tag{3.41}$$

Clearly, the larger L is, the larger the SMI, or the larger the uncertainty in the location of a particle within the range $(0, L)$. Note carefully that L has units of length. Therefore, the only meaningful application of (3.41) is for a *differences* in H. An example of a uniform distribution is shown in Figure 3.7.

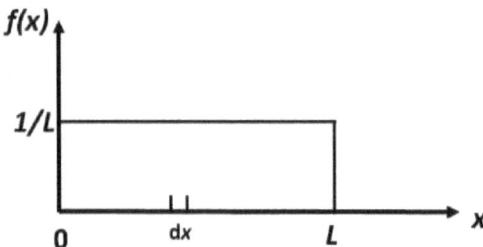

Figure 3.7. The uniform distribution of location.
The probability of finding a particle in a small interval dx,
between 0 and *L, is dx/L.*

In anticipating the application of this result in the next section, we divide the length L into n segments each of length h, i.e. $L = h \cdot n$ (h will later be the Planck constant, but here it is an arbitrary unit of length), Figure 3.8. We can use the property of consistency to express $H(0, L)$ as

$$H(0, L) = H\left[\frac{1}{n}, \cdots, \frac{1}{n}\right] + \sum_{i=1}^{n} \frac{1}{n} H[h] \tag{3.42}$$

This equation simply corresponds to rewriting the uncertainty regarding the location of the particle in the range $(0, L)$ in two terms; first, the uncertainty with respect to which of the n boxes of size h, and second, the average uncertainty in the location of the particle *within* the boxes of size h.

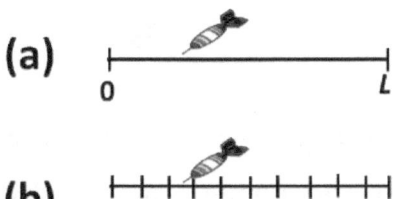

(a)

0   *L*

(b)

0   *L*

Figure 3.8. (a) A dart hits a one-dimensional segment of length L.
There are infinite possible locations for the dart.
(b) Passage from the infinite (continuum),
to the discrete description of the states.

From (3.41) and (3.42), we have

$$H(0, L) = \log n + \sum_{i=1}^{n} \frac{1}{n}\log h$$

$$= \log\frac{L}{h} + \log h = \log L \tag{3.43}$$

which is consistent with (3.41).

Now suppose that h is very small so that we are not concerned about the location within the box of size h. All we are concerned about is in which of the n boxes the particle is located. Clearly, the uncertainty in this case is simply the SMI of a discrete and finite case, i.e.

$$H\left[\frac{1}{n}, \cdots, \frac{1}{n}\right] = \log n = \log L - \log h \tag{3.44}$$

Thus, the subtraction of $\log h$ from $\log L$, amounts to the passage from the continuous to the discrete case of determining in which of the n boxes the particle is located. In practice, we can never determine the location of a particle with absolute or infinite accuracy. There is always a short interval of length wherein we cannot tell where the particle is. Hence, in all these cases, it is the discrete, rather than the continuous definition of H that applies. Note also that if we are not concerned about the location within the small intervals of length h, we get a pure number under the logarithm in (3.44).

The Normal distribution

The second theorem we present here is the following:

Of all the continuous distribution densities $f(x)$ for which the second moment (or the standard deviation) is finite and constant, the Gaussian (or the normal) distribution maximizes the SMI defined in eq. (3.31). The mathematical problem is to maximize H as defined in (3.31) subject to the two conditions

$$\int_{-\infty}^{\infty} f(x)dx = 1 \tag{3.45}$$

$$\int_{-\infty}^{\infty} x^2 f(x)dx = \sigma^2 (= constant) \tag{3.46}$$

We define the auxiliary functional

$$A[f(x)] = - \int f(x) \, \text{Log}[f(x)]dx + \lambda_1 \int f(x)x^2 dx + \lambda_2 \int f(x)dx \tag{3.47}$$

The functional derivative of $A[f(x)]$ with respect to the component $f(x')$ is:

$$\frac{\delta A}{\delta f(x\prime)} = -1 - \log f(x') + \lambda_1 x'^2 + \lambda_2 \tag{3.48}$$

The condition for maximum H is thus:

$$-1 - \log f^*(x) + \lambda_1 x^2 + \lambda_2 = 0 \tag{3.49}$$

The two Lagrange constants may be obtained by substituting (3.49) in (3.45) and (3.46) to obtain

$$f^*(x) = \exp[\lambda_1 x^2 + \lambda_2 - 1] \tag{3.50}$$

$$1 = \int_{-\infty}^{\infty} f^*(x)dx = \exp[\lambda_2 - 1] \int_{-\infty}^{\infty} \exp[\lambda_1 x^2]dx = \sqrt{-\frac{\pi}{\lambda_1}} \exp[\lambda_2 - 1] \tag{3.51}$$

$$\sigma^2 = \int_{-\infty}^{\infty} x^2 f^*(x)dx = \exp[\lambda_2 - 1] \int_{-\infty}^{\infty} x^2 \exp[\lambda_1 x^2]dx = \sqrt{-\frac{\pi}{4(-\lambda_1)^3}} \exp[\lambda_2 - 1] \tag{3.52}$$

From the last two equations, we can solve for λ_1 and λ_2, to obtain

$$\lambda_1 = \frac{-1}{2\sigma^2} \quad , \quad \exp[\lambda_2 - 1] = \frac{1}{\sqrt{2\pi\sigma^2}} \tag{3.53}$$

Hence, the equilibrium density is:

$$f^*(x) = f_{eq}(x) = \frac{\exp[-x^2/2\sigma^2]}{\sqrt{2\pi\sigma^2}} \tag{3.54}$$

The maximum value of the SMI (note that we use natural logarithm), is:

$$H_{max} = -\int_{-\infty}^{\infty} f^*(x) \log f^*(x)dx = \frac{1}{2}\log(2\pi e \sigma^2) \tag{3}$$

In the application of this result for the velocity distribution in one dimension, we have the probability distribution density:

$$f^*(v_x) = \sqrt{\frac{m}{2\pi k_B T}} \exp\left[\frac{-mv_x^2}{2k_B T}\right] \tag{3.56}$$

where we identify the standard deviation σ^2 as in Ben-Naim (2008, 2017b)

$$\sigma^2 = \frac{k_B T}{m} \tag{3.57}$$

where k_B is the Boltzmann constant, and T is the absolute temperature. In Figure 3.9 we show a few examples of the normal distribution.

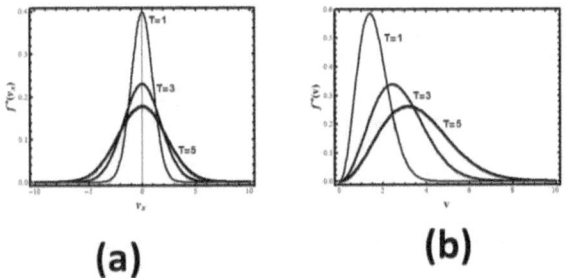

(a)   **(b)**

Figure 3.9. (a) The velocity distribution of particles in one dimension at different temperatures.
(b) The speed (or the absolute velocity) distribution of particles in 3D at different temperatures.

The SMI associated with the velocity distribution is thus

$$H(v_x) = \frac{1}{2}\log(2\pi e k_B T/m) \tag{3.58}$$

Similarly, for the momentum distribution in one dimension $p_x = mv_x$, we find that the density distribution which maximizes H is

$$f^*(p_x) = \frac{1}{\sqrt{2\pi m k_B T}} \exp\left[\frac{-p_x^2}{2m k_B T}\right] \tag{3.59}$$

and the corresponding maximum value of the SMI is

$$H_{max}(p_x) = \frac{1}{2}\log(2\pi e m k_B T) \tag{3.60}$$

In the next section, we shall use this last expression to construct the analogue of the Sackur-Tetrode equation for the entropy of an ideal gas.

The exponential, or the Boltzmann distribution

Another theorem proved by Shannon is that the Boltzmann distribution is obtained by maximizing H defined in (3.31), subject to the two conditions

$$\int_0^\infty f(x)dx = 1 \tag{3.61}$$

$$\int_0^\infty xf(x)dx = a \quad , \quad \text{with } a > 0 \tag{3.62}$$

Using the Lagrange multipliers λ_1 and λ_2, we seek the maximum of the functional

$$A[f(x)] = -\int_0^\infty f(x)\log f(x)dx + \lambda_1 \int_0^\infty f(x)dx + \lambda_2 \int_0^\infty xf(x)dx \tag{3.63}$$

The condition for an extremum is obtained by taking the functional derivative of $A[f(x)]$, and setting it equal to zero, i.e.

$$-1 - \log f^*(x) + \lambda_1 + \lambda_2 x = 0 \tag{3.64}$$

or equivalently

$$f^*(x) = \exp[\lambda_2 x + \lambda_1 - 1] \tag{3.65}$$

Substituting this density function in the two constraints (3.61) and (3.62), we get

$$\exp[\lambda_1 - 1] \int_0^\infty \exp[\lambda_2 x]\, dx = 1 \tag{3.66}$$

$$\exp[\lambda_1 - 1] \int_0^\infty x \exp[\lambda_2 x]\, dx = a \tag{3.67}$$

Note that λ_2 cannot be positive; otherwise, the two constraints cannot be satisfied (not even for $\lambda_2 = 0$). From (3.66) and (3.67), we obtain

$$\exp[\lambda_1 - 1] \left[\frac{\exp[\lambda_2 x]}{\lambda_2} \right]_0^\infty = -\frac{\exp[\lambda_1 - 1]}{\lambda_2} = 1 \tag{3.68}$$

and

$$\exp[\lambda_1 - 1] \left[\frac{(\lambda_2 x - 1) \exp[\lambda_2 x]}{\lambda_2^2} \right]_0^\infty = \frac{\exp[\lambda_1 - 1]}{\lambda_2^2} = a \tag{3.69}$$

These two equations can be solved for λ_1 and λ_2 to obtain

$$\lambda_2 = \frac{-1}{a} \tag{3.70}$$

$$\exp[\lambda_1 - 1] = \frac{1}{a} \tag{3.71}$$

Hence, the density function that maximized H is

$$\frac{1}{a}\exp\left(\frac{-x}{a}\right) \tag{3.72}$$

In Figure 3.10 we show a few examples of the exponential distribution for different values of a.

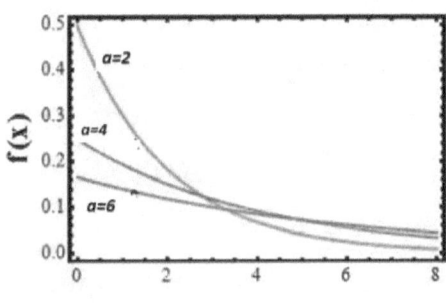

Figure 3.10. The exponential distribution for different values of a.

The value of the SMI corresponding to this density is (note that log is the natural logarithm).

$$H_{max} = \log a + 1 = \log(ae) \tag{3.73}$$

3.5 Derivation of the entropy of an ideal gas for simple particles

In this section, we *define* the concept of entropy as a special case of SMI. This definition leads naturally to a simple, intuitive and irrefutable interpretation of entropy. It also provides a solid probabilistic interpretation of the Second Law of Thermodynamics.

As was explained in section 3.1, Shannon *called* his measure, which we referred to as Shannon's Measure of Information (SMI), entropy. This was a huge mistake which caused great confusion in both information theory and in thermodynamics. The SMI is defined for any probability distribution. The entropy is defined on a tiny subset of all the possible distributions. Calling SMI entropy leads to many awkward statements such as: "The entropy tends to increase," or "The *value* of the maximum entropy at equilibrium is the *entropy* of the system."

The correct statement concerning the entropy of an isolated system is as follows: An isolated system at equilibrium is characterized by a fixed energy E, volume V and number of particles N (assuming a one-component system). For such a system, the entropy is *determined* by variables E, V, N (see below for a particular case of an explicit entropy function of an ideal gas). This entropy is fixed for that system. It is not a function of time, it does not change with time, and it does not tend to a maximum.

Similarly, one can define the entropy for any other well-defined thermodynamic system at equilibrium. This is exactly what is meant by the statement that entropy is a *state function*.

In this section we use for convenience the natural logarithm $\log_e x$, or $\ln x$. Whenever we want to convert to SMI we need to multiply by $\log_2 e$, i.e. $\log_2 x = \log_2 e \log_e x$.

The overall plan of obtaining the entropy of an ideal gas from the SMI consists of four steps:

First, we calculate the *locational* SMI associated with the *equilibrium* distribution of locations of all the particles in the system.

Second, we calculate the *velocity* SMI associated with the *equilibrium* distribution of velocities (or momenta) of all the particles.

Third, we add a correction term due to the quantum mechanical *uncertainty principle*.

Fourth, we add a correction term due to the fact that the particles are *indistinguishable*.

Note that in the first two steps, we use the maximum SMI method to find out the *equilibrium distribution*. Then we use this equilibrium distribution to *evaluate* the corresponding SMI. In the last two steps we introduce two corrections to the SMI. Once we combine the results of the four steps, we get, up to a multiplicative constant, the *entropy* of an ideal gas.

3.5.1 The locational SMI of a particle in a 1D box of length L

Suppose we have a particle confined to a one-dimensional (1D) "box" of length L. As we discussed in Section 3.4, since there are infinite points in which the particle can be within the interval $(0, L)$, the corresponding locational SMI must be infinite. However, we can define, as Shannon did, the following quantity by analogy with the discrete case:

$$H(X) = - \int f(x) \log f(x) dx \tag{3.74}$$

This quantity might either converge or diverge, but in any case, in practice we shall use only differences between such quantities. In section 3.4 we calculated the density distribution which maximizes the locational SMI, $H(X)$ in (3.74) and found

$$f_{eq}(x) = \frac{1}{L} \qquad (3.75)$$

The use of the subscript *eq.* (for equilibrium) will be clarified later. The corresponding SMI calculated by substituting (3.75) in (3.74) is:

$$H(\text{locations in } 1D) = \log L \qquad (3.76)$$

We now acknowledge that the location of the particle cannot be determined with absolute accuracy, i.e. there exists a small interval h_x within which we do not care where the particle is. Therefore, we must correct equation (3.76) by subtracting $\log h_x$. Thus, we write instead of (3.76)

$$H(X) = \log L - \log h_x \qquad (3.77)$$

We recognize that in (3.77) we effectively defined $H(X)$ for the *finite* number of intervals $n = L/h$. See Figure 3.8. Note that when $h_x \to 0$, $H(X)$ diverges to infinity. Here, we do not take the mathematical limit, but we stop at h_x small enough but not zero. Note also that in writing eq. (3.77) we do not have to specify the units of length, as long as we use the same units for L and h_x.

3.5.2 The velocity SMI of a particle in a 1D "box" of length L

In section 3.4, we calculated the probability distribution that maximizes the continuous SMI, subject to two conditions

$$\int_{-\infty}^{\infty} f(x)dx = 1 \qquad (3.78)$$

$$\int_{-\infty}^{\infty} x^2 f(x)dx = \sigma^2 = constant \qquad (3.79)$$

The result is the Normal distribution

$$f_{eq}(x) = \frac{\exp[-x^2/2\sigma^2]}{\sqrt{2\pi\sigma^2}} \tag{3.80}$$

The subscript eq which stands for equilibrium will be clarified once we realize that this is the equilibrium distribution of velocities. Applying this result to a classical particle having average kinetic energy $\frac{m<v_x^2>}{2}$, and using the relationship between the standard deviation σ^2 and the temperature of the system

$$\sigma^2 = \frac{k_B T}{m} \tag{3.81}$$

We get the equilibrium velocity distribution of one particle in a 1D system

$$f_{eq}(v_x) = \sqrt{\frac{m}{2\pi k_B T}} \exp\left[\frac{-mv_x^2}{2k_B T}\right] \tag{3.82}$$

where k_B is the Boltzmann constant, m is the mass of the particle, and T the absolute temperature. The value of the continuous SMI for this probability density is

$$H_{max}(velocity\ in\ 1D) = \frac{1}{2}\log(2\pi e k_B T/m) \tag{3.83}$$

Similarly, we can write the momentum distribution in 1D, by transforming from $v_x \rightarrow p_x = mv_x$, to get

$$f_{eq}(p_x) = \frac{1}{\sqrt{2\pi m k_B T}} \exp\left[\frac{-p_x^2}{2m k_B T}\right] \tag{3.84}$$

and the corresponding maximum SMI

$$H_{max}(momentum\ in\ 1\ D) = \frac{1}{2}\log(2\pi emk_BT) \qquad (3.85)$$

As we have noted in connection with the locational SMI, the quantities (3.83) and (3.85) were calculated using the definition of the *continuous* SMI. Again, recognizing the fact that there is a limit to the accuracy within which we can determine the velocity (or the momentum) of the particle, we correct the expression in (3.85) by subtracting $\log h_p$ where h_p is a small, but finite interval.

$$H_{max}(momentum\ in\ 1D) = \frac{1}{2}\log(2\pi emk_BT) - \log h_p \qquad (3.86)$$

Note again that if we choose the units of h_p of momentum as, *mass length/time*, the same as of $\sqrt{mk_BT}$, then the whole expression under the logarithm, will be a pure number.

3.5.3 Combining the SMI for the location and momentum of one particle in 1D system

In the previous two sections, we derived the expressions for the locational and the momentum SMI of one particle in 1D system. We now combine the two results. Assuming that the location and the momentum (or velocity) of the particles are independent events we write

$$H_{max}(location\ and\ momentum) = H_{max}(location) +$$

$$H_{max}(momentum) = \log\left[\frac{L\sqrt{2\pi emk_BT}}{h_x h_p}\right] \qquad (3.87)$$

Recall that h_x and h_p were chosen to eliminate the divergence of the SMI. In writing (3.87) we assume that the location and the momentum of the particle are independent. However, quantum mechanics impose restrictions on the accuracy in determining both the location x and the corresponding momentum p_x. In equations (3.77) and (3.85), h_x and h_p were introduced because we did not care to determine the location and the momentum with an accuracy better than h_x and h_p, respectively. Now, we must acknowledge that nature imposes on us a limit on the accuracy with which we can determine *simultaneously* the location and the corresponding momentum. Thus, in equation (3.87), h_x and h_p cannot both be arbitrarily small, but their product must be of the order of Planck constant $h = 6.626 \times 10^{-34} J s$. Thus, we set:

$$h_x h_p \approx h \qquad (3.88)$$

And instead of (3.87), we write

$$H_{max}(location\ and\ momentum) = \log\left[\frac{L\sqrt{2\pi e m k_B T}}{h}\right] \qquad (3.89)$$

3.5.4 The SMI of a particle in a box of volume V

We consider again one simple particle in a cubic box of volume V. We assume that the location of the particle along the three axes x, y and z are independent. Therefore, we can write the SMI of the location of the particle in a cube of edges L, and volume V a

$$H(location\ in\ 3D) = 3H_{max}(location\ in\ 1D) = 3 \log L = \log V \ (3.90)$$

Similarly, for the momentum of the particle we assume that the momentum (or the velocity) along the three axes x, y and z are independent. Hence, we write:

$$H_{max}(momentum\ in\ 3D) = 3H_{max}(momentum\ in\ 1D) \qquad (3.91)$$

We combine the SMI of the locations and momenta of one particle in a box of volume V, taking into account the uncertainty principle. The result is:

$$H_{max}(location\ and\ momentum\ in\ 3D) =$$

$$3 \log[\frac{L\sqrt{2\pi emk_BT}}{h}] \qquad (3.92)$$

3.5.5 The SMI of locations and momenta of N independent particles in a box of volume V

The next step is to proceed from one particle in a box to N independent particles in a box of volume V. Given the location (x, y, z), and the momentum (p_x, p_y, p_z) of one particle within the box, we say that we know the *microstate* of the particle. If there are N particles in the box, and if their microstates are independent, we can write the SMI of N such particles simply as N times the SMI of one particle, i.e.

$$SMI(\ N\ independent\ particles) = N \times SMI(one\ particle) \quad (3.93)$$

This equation would have been correct if the microstates of all the particles were independent. In reality, there are always correlations between the microstates of all the particles; one is due to the *indistinguishability* between the particles. The second is due to *intermolecular interactions* between the particles. We shall discuss these two sources of correlations separately.

(i) correlation due to indistinguishability

Recall that the microstate of a single particle includes the location and the momentum of that particle. Let us focus on the location of one particle in a box of volume V. We have written the locational SMI in eq. (3.90) as:

$$H_{max}(location) = 3 \log L = \log V \qquad (3.94)$$

Recall that this result was obtained for the continuous locational SMI. This result does not take into account the divergence of the limiting procedure. In order to explain the source of the correlation due to indistinguishability, we divide the volume V into a very large number M, of small cells each of the volume V/M. We are not interested in the exact location of each particle, but only in which of the cells each particle is located. The total number of cells is M, and we assume that the total number of particles is very small compared with M, $N \ll M$. A simple case is: $N = 2$ and $M = 100$, as shown in Figure 3.11. If each cell can contain at most one particle, then there are M possibilities as to where to put the first particle in one of the cells, and there are $M - 1$ possibilities as to where to put the second particle in the remaining empty cells. Altogether, we have $M(M - 1)$ possible microstates, or configurations for the two particles.

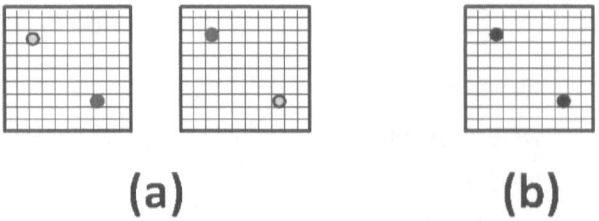

(a)   **(b)**

Figure 3.11. Two different particles in 100 cells (N=2, M=100).
(a) Two different configurations become one configuration,
(b) When the particles are indistinguishable

The probability that a particle is found in cell j *(or cell i)* is:

$$\Pr(j) = \Pr(i) = \frac{1}{M} \tag{3.95}$$

Because the two particles cannot be in the same cell, the probability that one particle is found in cell i, and the second in a different cell j is:

$$\Pr(i,j) = \frac{1}{M(M-1)} \tag{3.96}$$

Thus, we see that even in this simple example, there is a *correlation* between the events "one particle in i," and one particle in j." The correlation function measures the extent of dependence between the two events:

$$g(i,j) = \frac{\Pr(i,j)}{\Pr(i)\,\Pr(j)} = \frac{M^2}{M(M-1)} = \frac{1}{1-\frac{1}{M}} \tag{3.97}$$

This correlation is easy to understand. For any finite M, the conditional probability of finding a particle in cell i given another particle in cell j, is different from the (unconditional) probability of finding the particle in cell i.

Clearly, this correlation can be made as small as we like, by taking $M \gg 1$ (or in general, $M \gg N$). There is another correlation which we cannot eliminate and is due to the indistinguishability of the particles.

Note that in counting the total number of configurations we have used the two labeled particles, say, white and black. In this case we count the two configurations in Figure 3.11a as *different* configurations: "black particle in cell i, and white particle in cell j," and "black particle in cell j, and white particle in cell i."

Atoms and molecules are indistinguishable by nature; we cannot label them. Therefore, the two microstates (or configurations) in Figure 3.11a are *indistinguishable*. This means that the total number of configurations is not $M(M - 1)$, but

$$\frac{M(M-1)}{2} \rightarrow \frac{M^2}{2} \quad \text{(for large } M\text{)} \tag{3.98}$$

For very large M we have a correlation between the events "particle in i," and "particle in j."

$$g(i,j) = \frac{\Pr(i,j)}{\Pr(i)\,\Pr(j)} = \frac{M^2}{M^2/2} = 2 \tag{3.99}$$

For N particles distributed in M cells, we have a correlation function (For $M \gg N$)

$$g(i_1, i_2, \ldots, i_n) = \frac{M^N}{M^N/N!} = N! \tag{3.100}$$

This means that for N indistinguishable particles we must divide the number of configurations M^N by $N!$. Thus, in general by removing the "labels" on the particles the number of configurations is *reduced* by the factor $N!$ For two particles the two configurations shown in Figure 3.11a reduce to one shown in Figure 3.11b.

Now that we know that there are correlations between the events "one particle in i_1," "one particle in i_2," ... "one particle in i_n", we can define the *mutual information* corresponding to this correlation. We write this as

$$I(1; 2; ...; N) = \log N! \qquad (3.101)$$

The SMI for N indistinguishable particles will then be

$$H(N \ particles) = \sum_{i=1}^{N} H(one \ particle) - \log N! \qquad (3.102)$$

For the definition of the total mutual information, see Ben-Naim (2017c).

Using the SMI for the location and momentum of one particle in (3.92) we can write the final result for the SMI of N indistinguishable (but non-interacting) particles as:

$$H(N \ indistinguishable \ particles)$$

$$= N\log V \left(\frac{2\pi m e k_B T}{h^2}\right)^{3/2} - \log N! \qquad (3.103)$$

Using the Stirling approximation for $\log N!$ (note again that we use here the natural logarithm) in the form:

$$\log N! \approx N\log N - N \qquad (3.104)$$

We have the final result for the SMI of N indistinguishable particles in a box of volume V, and temperature T:

$$H(1,2,...N) = N\log\left[\frac{V}{N}\left(\frac{2\pi m k_B T}{h^2}\right)^{3/2}\right] + \frac{5}{2}N \qquad (3.105)$$

This is a remarkable result. By multiplying the SMI of N particles in a box of volume V, at temperature T, by a constant factor (k_B, if we use the natural log, or $k_B \log_e 2$ if the log is to the base 2), one gets the *entropy*, the *thermodynamic entropy* of an ideal gas of simple particles. This equation was derived by Sackur and by Tetrode in 1912, by using the Boltzmann definition of entropy. Here, we have derived the entropy function of an ideal gas from the SMI. See also Ben-Naim (2008)

One can convert this expression to the *entropy function* $S(E,V,N)$, by using the relationship between the total kinetic energy of the system, and the total kinetic energy of all the particles:

$$E = N\frac{m\langle v\rangle^2}{2} = \frac{3}{2}Nk_B T \qquad (3.106)$$

The explicit entropy function of an ideal gas is obtained from (3.105) and (3.106):

$$S(E,V,N) = Nk_B \ln\left[\frac{V}{N}\left(\frac{E}{N}\right)^{3/2}\right] + \frac{3}{2}k_B N\left[\frac{5}{3} + \ln\left(\frac{4\pi m}{3h^2}\right)\right] \qquad (3.107)$$

We can use this equation as a *definition* of the entropy of an ideal gas of simple particles characterized by constant energy, volume and number of particles. Note that when we combine all the terms under the logarithm sign, we must get a dimensionless quantity. For more details, see Ben-Naim (2017).

The next step in the definition of entropy is to add to the entropy of an ideal gas, the *mutual information* due to intermolecular interactions.

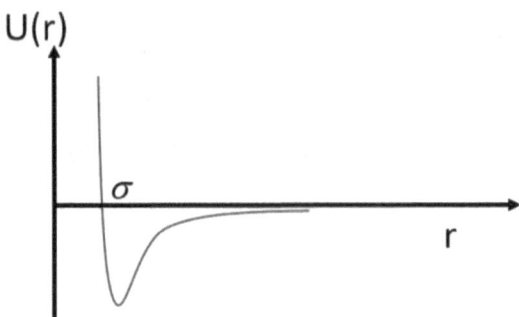

Figure 3.12. Typical form of a pair potential between two simple particles

(ii)   Correlation due to intermolecular interaction

In equations (3.105) or (3.107) we got the entropy of a system of non-interacting simple particles (ideal gas). In any real system of particles, there are some interactions between the particles. One example of an energy potential function, *U(r),* is shown in Figure 3.12. Without going into details on the function $U(r)$ shown in the figure, it is clear that there are two regions of distances $0 \leq r \lesssim \sigma$ and $\sigma \leq r \lesssim \infty$, where the slope of the function $U(r)$ is negative and positive, respectively. Negative slope corresponds to repulsive forces between the pair of the particles when they are at a distance smaller than σ. This is the reason why σ is sometimes referred to as the *effective diameter* of the particles, not to be confused with the variance in eq. (3.81). For bigger distances, $r \gtrsim \sigma$ we observe attractive forces between the particles.

Intuitively, it is clear that interactions between the particles induce correlations between the locational probabilities of the two particles. For hard-sphere particles there is an infinitely strong repulsive force between two particles when they approach a distance of $r \leq \sigma$. Thus, if we know the location R_1 of one particle, we can be sure that a second particle, at R_2, is not in a sphere of diameter σ around the point R_1. The distance between R_1 and R_2 is $r=|R_2 - R_1|$. This *repulsive* interaction is said to introduce *negative correlation* between the locations of the two particles, i.e. g <1.

On the other hand, two argon atoms *attract* each other at distances $r >$ 4Å. Therefore, if we know the location of one particle say, at R_1, the probability of observing a second particle at R_2 is bigger than the probability of finding the particle at R_2 in the absence of a particle at R_1. In this case we get *positive correlation* between the locations of the two particles, i.e. g >1.

We can conclude that in both cases (attraction and repulsion) there are correlations between the particles. These correlations can be cast in the form of mutual information which *reduces* the SMI of a system of N particles in an ideal gas phase. The mathematical details of these correlations are discussed in Ben-Naim (2008, 2017b).

Here, we show only the form of the mutual information for the limit of very low densities. At this limit, we can assume that there are only *pair correlations*, and neglect all higher order correlations. The mutual information due to these correlations has the form:

$I(due\ to\ correlations\ in\ pairs)=$

$$\frac{N(N-1)}{2} \int \Pr(\boldsymbol{R_1}, \boldsymbol{R_2}) \log g(\boldsymbol{R_1}, \boldsymbol{R_2}) d\boldsymbol{R_1} d\boldsymbol{R_2} \qquad (3.108)$$

where $\Pr(\boldsymbol{R_1}, \boldsymbol{R_2}) d\boldsymbol{R_1} d\boldsymbol{R_2}$ is the probability of finding one particle in $d\boldsymbol{R_1}$ at $\boldsymbol{R_1}$ and a second particle in $d\boldsymbol{R_2}$ at $\boldsymbol{R_2}$. The correlation function is defined by (see also section 4.5):

$$g(\boldsymbol{R_1}, \boldsymbol{R_2}) = \frac{\Pr(\boldsymbol{R_1}, \boldsymbol{R_2})}{\Pr(\boldsymbol{R_1})\Pr(\boldsymbol{R_2})} \qquad (3.109)$$

Note again that log g can be either positive or negative, but the average in (3.108) must be positive.

3.5.6 Conclusion

We summarize the main steps leading from the SMI to the entropy.

We started with the SMI associated with the *locations* and *momenta* of the particles. We calculated the distribution of the locations and momenta that *maximizes* the SMI. We referred to this distribution as the *equilibrium distribution*. Let us denote this distribution of the locations and momenta of all the particles by $f_{eq}(\boldsymbol{R}, \boldsymbol{p})$.

Next, we use the equilibrium distribution to calculate the SMI of a system of N particles in a volume V, and at temperature T. This SMI is, up to a multiplicative constant ($k_B \ln 2$), identical to the *entropy* of an ideal gas at *equilibrium*. This is the reason we referred to the distribution which maximizes the SMI (denoted f^* in section 3.4) as the *equilibrium distribution*.

It should be noted that in the derivation of the entropy, we used the SMI twice; first, in calculating the distribution that maximizes the SMI, then in evaluating the maximum SMI corresponding to this distribution. The distinction between the concepts of SMI and entropy is essential. Referring to the SMI (as many do) as entropy, inevitably leads to such an awkward statement; the maximum value of the entropy (meaning the SMI) is the entropy (meaning the thermodynamic entropy). The correct statement is that the SMI associated with locations and momenta is defined for any system; small or large, at equilibrium or far from equilibrium. This SMI, not the entropy, evolves into a maximum value when the system reaches equilibrium. At this state, the SMI becomes proportional to the entropy of the system. The entropy obtained in this procedure is referred to as the Shannon's based or the ABN definition of entropy.

Since the entropy is, up to a constant, a special case of an SMI, it follows that whatever interpretation one accepts for the SMI, it will be automatically applied to the concept of entropy. The most important conclusion of this definition is that entropy, being a state function, *is not a function of time*. Entropy does not change with time, and entropy does not have a tendency to increase.

We can summarize the ABN *definition* of entropy as follows. We start with the probability distribution of the locations and momenta of all the particles. On this distribution we define the SMI, then we take the maximum value of the SMI, and add two corrections to get the entropy of an ideal gas at *equilibrium*. One can easily extend this definition to a system of interacting particles by adding the appropriate mutual information. Additionally, it is easy to add any entropy due to internal degrees of freedom.

From this definition it is clear that entropy is defined for a well-defined thermodynamic system at equilibrium. It is also clear that entropy is not defined for any living system, and not for the entire universe. Unfortunately, many recent popular science books, invoke the concept of entropy to "explain" many unexplainable phenomena associated with life. No doubt, this practice has contributed to make entropy the most mysterious concept in science. For more on this see Ben-Naim (2015a, 2016a, 2016b)

Chapter 4. The Second Law of Thermodynamics

In most textbooks on thermodynamics, the concepts of entropy, and the Second Law are intertwined. This is so in Clausius' statement that "the entropy of the universe always increases" to a more recent statement by Atkins (2007): "The Second Law: The increase in entropy."

In the previous chapter we have introduced the concept of entropy without mentioning the Second Law. Likewise, in this chapter we show that the Second Law may be formulated without mentioning the entropy. It is true that for some specific processes in an isolated system the Second Law may be formulated in terms of entropy. However, this is not the most general formulation of the Second Law. In fact, the entropy-formulation of the Second Law might even be misleading. Some people see entropy change as the *cause* of a spontaneous process, not as a *result* of the process.

In this chapter we start with a few simple processes of "expansion" with small number of particles. We discuss the meaning of irreversibility in these processes. We shall see that only in the limit of very large number of particles, can the process be considered to be irreversible.

In the derivation of the entropy function in Chapter 3, we used the maximum SMI with respect to *molecular* probability distributions. In this chapter we formulate the Second Law in terms of maximum entropy over all possible *constrained equilibrium states*. Before doing this we should clarify the meaning of the two terms: Maximum entropy (MaxEnt), and MaxSMI.

Ever since Jaynes introduced the principle of MaxEnt (following Shannon's naming of the SMI as entropy), people actually used the principle of MaxSMI but referred to it as MaxEnt. This is potentially confusing. The entropy is obtained (up to a constant) as a MaxSMI. We shall soon state the entropy-formulation of the Second Law, where we use the term of Max Entropy over all possible entropies of the same system but at *constrained equilibrium states*, Figure 4.1. At these equilibrium states the entropy is defined for each subsystem. Each of these entropies are related to the maximum of the SMI (not of entropy) of the locational and momentum distribution in that subsystem, as defined in the previous chapter.

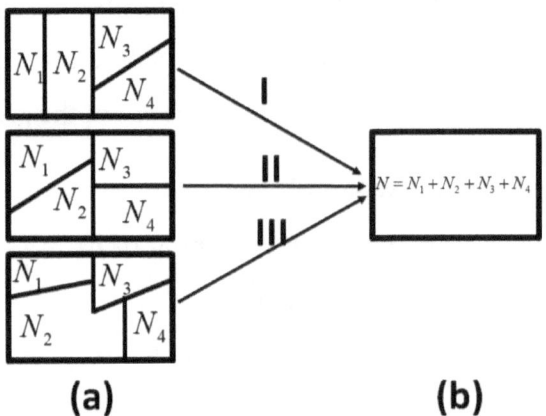

(a)   (b)

Figure 4.1. (a) Three different constrained equilibrium systems, all having the same total E, V, and N. (b) The unconstrained equilibrium state obtained after the removal of the constraints.

It is important to distinguish between the thermodynamics Max Entropy and the molecular MaxSMI.

For a given system at any distribution of locations and momenta, one can define the corresponding SMI. This SMI is not entropy. Only when the SMI reaches its maximal value is it related to the entropy of the system.

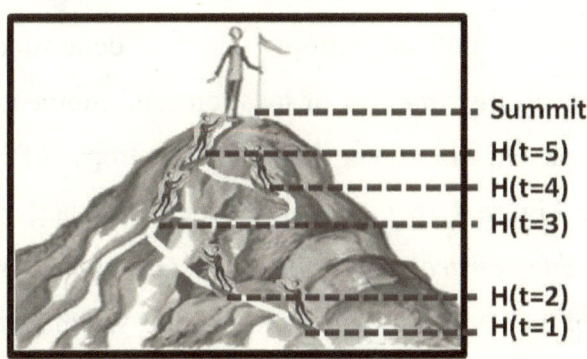

Figure 4.2. A person climbing a hill.
The height at which the person is, changes with time, H(t),
but the summit does not change with time.

To highlight the difference between the SMI and entropy consider a person climbing up a hill, Figure 4.2. The height, H, at which the person is, above some reference level, is the analogue of the SMI. It can change with time depending on the route and the walking speed. Upon reaching the summit (the analogue of the entropy), the height at which the person is, can either stay constant or decrease, but it cannot increase since by definition the summit is the maximum height of the hill. It would be awkward to say that the summit of the hill changes as the person walks, and that the summit increases towards the peak (which is the summit).

By the same token, it would be incorrect to say that the entropy of a system increases to its maximum value which is the entropy of the system. The correct statement is that the SMI (or the height) increases in the process until it reaches its maximum value (the summit). At that point the maximum SMI is related to the entropy of the system.

Note that in Figure 4.1a each subsystem is a macroscopic system at equilibrium having a well-defined entropy, which is determined by the maximum SMI (over all distributions of locations and momenta) in that particular subsystem. On the other hand, the entropy of the final equilibrium state in Figure 4.1b is, by the Second Law larger than the entropy of any *possible constrained equilibrium state of the system*. It is also the maximum of the SMI of the final state of the system, Figure 4.1b.

Clearly, each constrained equilibrium state *defines* a locational molecular distribution. However, not every molecular locational distribution corresponds a constrained equilibrium state.

Similarly, for each constrained equilibrium state, with different temperatures $T_1, T_2, \dots, T_c$, corresponds to a molecular velocity distribution of the entire system. The converse of this statement is not true. It is not true that any distribution of velocities corresponds to a constrained equilibrium state. For a simple example, see Ben-Naim (2017c), "The Four Law That Do Not Drive the Universe."

4.1 What drives the system to an equilibrium state?

We shall answer the question posed in the title of this section by examining a simple example. This example leads to an important conclusion regarding the question of reversibility in thermodynamic processes.

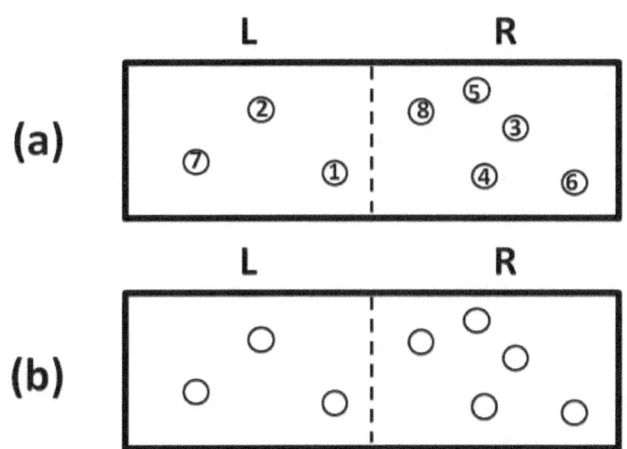

Figure 4.3. (a) A *specific* configuration of eight particles in the two compartments. Here, particles 1, 2 and 7 are in L, and 3,4,5,6, and 8 are in R. (b) A generic configuration, where we have three particles in L and five in R.

Consider a system of N non-interacting particles (ideal gas) in a volume $2V$ at constant energy E. We divide the system into two compartments L and R, each of volume V, Figure 4.3. We define the *specific* description of the state of the system when we are given $E, 2V, N$, and in addition we know which *specific* particles are in the right compartment (R), and which specific particles are in the left compartment (L). The *generic* or the macroscopic description of the same system is $(E, 2V, N; n)$ where n is the *number* of particles in the compartment L. Thus, in the specific description we are given a *specific* configuration of the system as if the particles were labeled, $1, 2, \cdots, N$. Here, by *configuration* we mean only which particles are in R and which are in L. In the *generic* description, we are given the information only on the *number* of particles in each compartment.

Clearly, if we only know that there are n particles in L, and $N - n$ particles are in R, we have many *specific* configurations that are consistent with the requirement that there are n particles in L.

We denote by $W(n)$ the number of specific configurations consistent with n particles in L. The first postulate of statistical mechanics states that *all specific* configurations of the system are equally probable. Clearly, the total number of *specific* configurations is 2^N, i.e. each particle can be in either one of the two compartments. Using the *classical* definition of the probability, we can calculate the probability of finding n particles in L and $(N - n)$ particles in R. We denote this probability by $P_N(n)$. It is easy to show that both $W(n)$ and $P_N(n)$ have a maximum as a function of n at the point $n^* = \frac{N}{2}$. (See below). The maximum value of the probability $P_N(n)$ (obtained at $n^* = \frac{N}{2}$), and is denoted by $P_N(n^*)$.

Thus, for any given N, there exists an n, such that the number of configurations, $W(n)$, or of the probability, $P_N(n)$ is maximal. Therefore, if we prepare a system with any initial distribution of particles n, and $N - n$ in the two compartments, and let the system evolve, the system's state will change from a state of lower probability to a higher probability. As N increases, the value of the maximum number of configurations $W(n^*)$

increases with N. However, the value of the maximal probability $P_N(n^*)$ *decreases* with N.

To appreciate the significance of this fact, we will examine the "evolution" of systems having small numbers of particles. We shall see in what sense the spontaneous process of expansion proceeds in "one direction only," or is "irreversible." Later, we shall also follow the changes in the SMI in the process of expansion and finally, we shall calculate the entropy change for this process. For some simulations, see Ben-Naim (2008, 2010).

In the following calculation we shall use the distribution, (p, q) *where:* $p = n/N$, and $q = 1 - p = (N - n)/N$. On this distribution we calculate the Probability (Pr) and the SMI associated with this distribution.

The case of two particles: $N = 2$

Suppose we have the total of $N = 2$ particles, Figure 4.4. In this case, we have the following possible *generic* configurations and the corresponding probabilities:

$n = 0$ $n = 1$ $n = 2,$

$$P_N(0) = \frac{1}{4}, \quad P_N(1) = \frac{1}{2}, \quad P_N(2) = \frac{1}{4} \qquad (4.1)$$

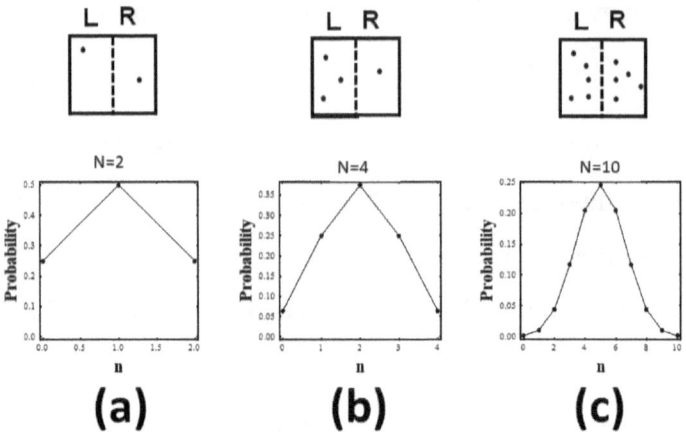

Figure 4.4. The probability of observing *n* particles in one compartment, and *N-n* in the other, for different numbers *N:* N=2, N=4 and N=10

This means that on the average, we can expect to find the configuration $n = 1$ (i.e. one particle in each compartment) about half of the time, but each of the configurations $n = 0$ and $n = 2$ only a quarter of the time, Figure 4.4a. If we start with all the particles in the left compartment, we shall find that the system will "expand" from V to $2V$. However, once in a while the two particles will be found in one compartment.

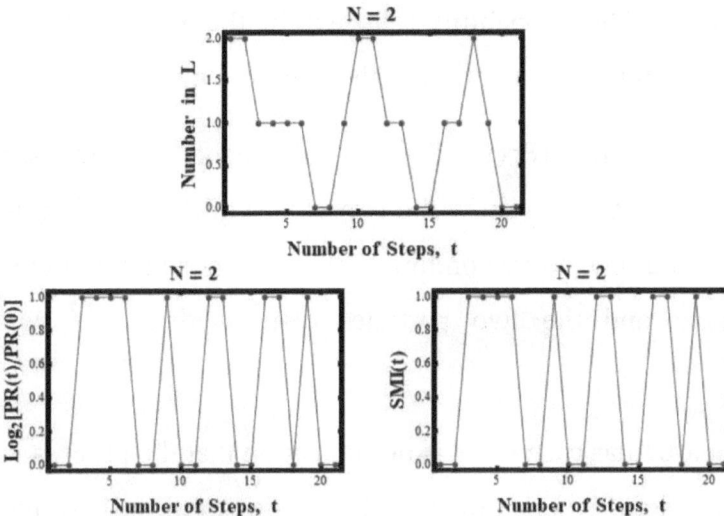

Figure 4.5. Simulated results for *N=2*.

Figure 4.5 shows some simulated results for this experiment. We start with all the particles in the left compartment (L). We then choose a particle at random and transfer it to a random compartment, in this case to either L or R. More details on the program of the simulations with different numbers of particles, and different numbers of cells may be found in Ben-Naim (2010).

We see that the number of particles in L starts with 2, then the number fluctuates between 0, 1, 2. In this particular run we find about 4 snapshots in which there are 2 in L, and 0 in R, about 6 with 0 in L, and 2 in L, and 10 with 1 in L, 1 in R. There are two probabilities in this case $\frac{1}{4}$ and $\frac{1}{2}$. We plot in Fig. 4.5 $\log_2[\Pr(t)/\Pr(0)]$, i.e. the probability of the state at step t, relative to the probability of the initial state (when all particles are in L). We plot the $\log_2$ of this ratio because in the next few cases the probability ratio will be very large. This probability ratio will be the basis on which we shall state the probability formulation of the Second Law.

The two values of $\log_2[\text{Pr}(t)/\text{Pr}(0)]$ correspond to the two probabilities: ¼ and ½ . We also show the two values of the SMI in this case which are zero and one, corresponding to the two cases; all particles in one compartment and the two particles distributed in the two compartments.

Clearly, in this particular case there is no indication that the process is "irreversible" or that there is a one-directional evolution of the state of the system.

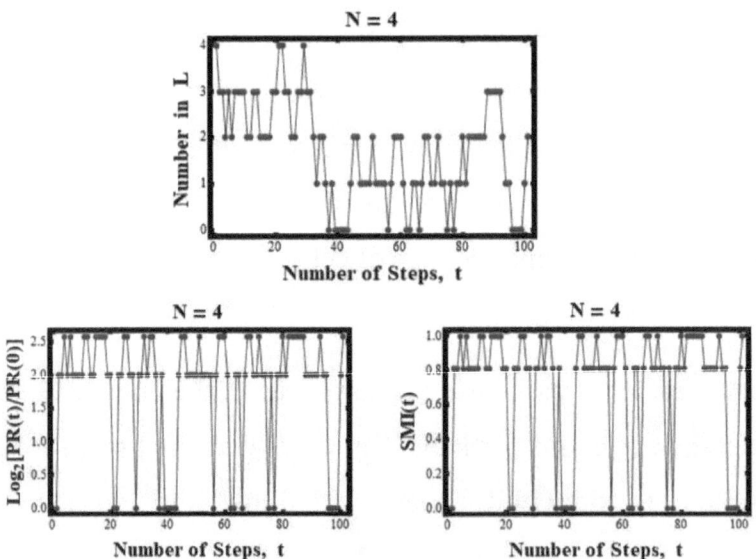

Figure 4.6. Simulated results for *N=4*.

The case of $N = 4$ is shown in Figures 4.4 and 4.6. Here, we see fewer visits to the initial state as well as to the state with zero particles in L. Most of the states are such that there are two particles in each compartment.

In this case there are only three values for the probability ratio $Pr(t)/Pr(0)$. These are:

$$1 \qquad\qquad \frac{4}{16}/\frac{1}{16} = 4 \qquad\qquad \frac{6}{16}/\frac{1}{16} = \frac{6}{1} = 6$$

$$\log_2 1 = 0 \qquad \log_2 4 = 2 \qquad\qquad \log_2 6 \approx 2.58 \qquad\qquad (4.2)$$

Corresponding to these three states we have three values of the SMI (per particle) 0, 0.81 and 1.

Again, there is no clear "one-directional" evolution of the state of the system. The system does return to the initial state, and the value of the SMI does return to the initial value of zero.

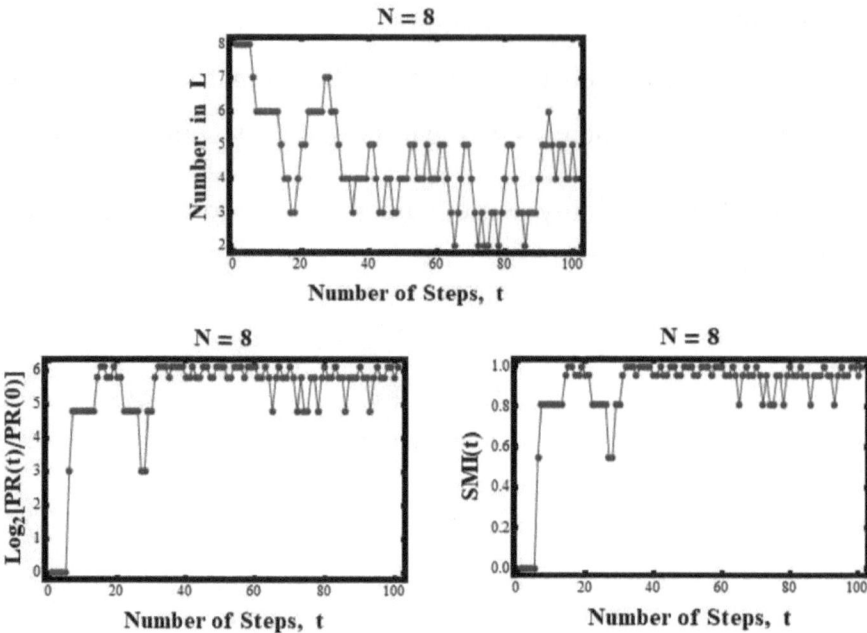

Figure 4.7. Simulated results for *N=8*.

In Figure 4.7 we show the result with $N = 8$. In this particular case, we do not see any single visit to the initial state after 100 steps. Of course, for a longer run we shall find the initial state in about one in $2^8 = 256$ steps or snapshots. The SMI also does not go back to the initial value. We see noticeable fluctuations in the value of the SMI with a maximum value again of 1 (this is the SMI per particle, see below).

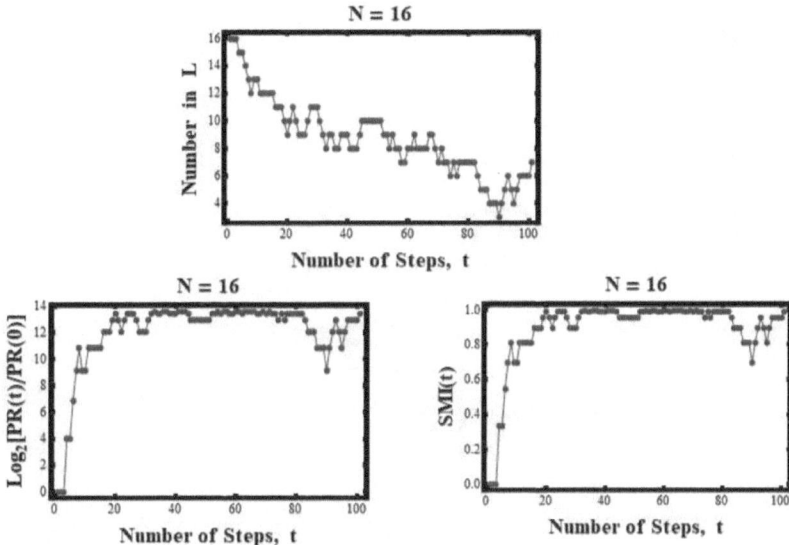

Figure 4.8. Simulated results for *N=16*.

For the case $N = 16$, we see that the number of particles initially drops from the value of 16 to about 8, then fluctuate around this value, Figure 4.8. On the other hand, the SMI starts at 0 then climbs steadily (though not strictly monotonically) to about 1. After about 60 steps the value of SMI is almost always at its maximal value of 1. There are some small fluctuations but these are relatively rare. Also, note that the maximal probability ratio is already very large of the order 2^{30} which is a huge number. This means that after about 60 steps the state of the system already reaches the one with the highest probability.

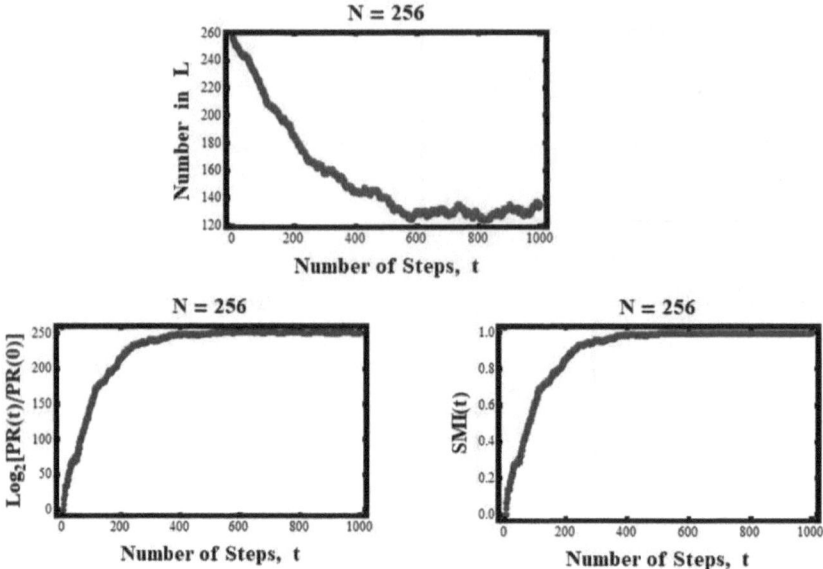

Figure 4.9. Simulated results for *N=256*.

For $2^8 = 256$ particles we see that the number of particles in L steadily decreases from the initial value of 256 to about 128 then fluctuates about this number, Figure 4.9. Also, the probability ratio climbs almost monotonically to its maximal value which is a huge number of the order 2^{256}.

Similarly, the SMI per particle climbs steadily from the initial value of 0 to the maximal value of 1.

Note that the fluctuations of the number of particles in L is about the value of $256/2 = 128$. The fluctuation in the value of SMI is about a value slightly smaller than 1.

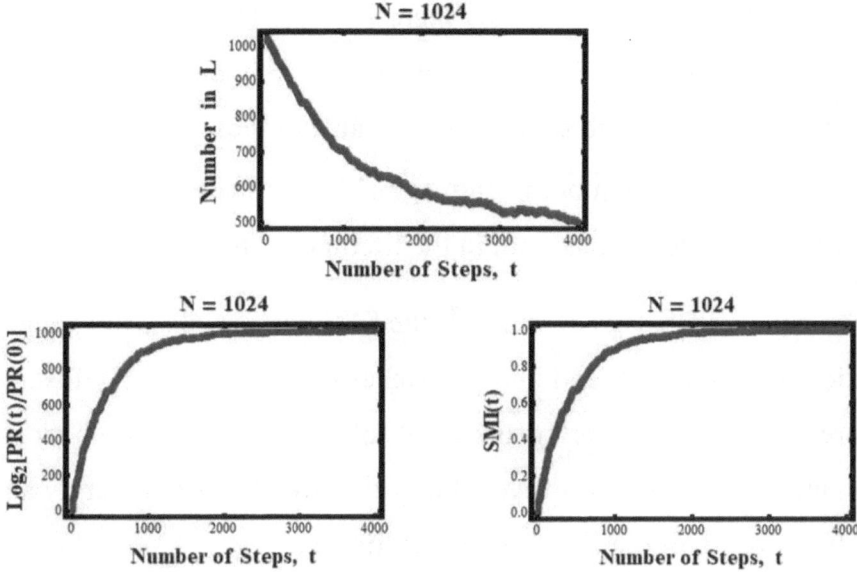

Figure 4.10. Simulated results for *N=1024*.

Finally, we show in Figure 4.10 the results for the case of $2^{10} = 1024$. Again, we notice that the curves became smoother, the ascent of the SMI nearly monotonic, and once it reaches the maximal value it stays there for a long time. While it is clear that in this case no return to the initial state is observed, and no significant fluctuations in the SMI are observed, we cannot conclude that the process is *strictly irreversible*. If we wait long enough we shall observe visits to the initial state, as well as a decrease in the SMI to its initial state. Such events are not *impossible,* but *extremely improbable*. They could occur in about 1 in 2^{1024} snapshots which is a huge number, about 10^{300}, i.e. 10 followed by 300 zeros!

One can imagine that for N of the order of 10^{23} the curves we have seen in Figures 4.7 to 4.10 will become almost monotonic, and once we reach the maximal value, the SMI will stay there forever. This "forever" does not mean that the process is *strictly irreversible*, it only means that reversing the value of SMI to say, the initial value of 0 can occur in about $2^{10^{23}}$ snapshots. Such a number is unimaginable; we shall not see such a reversal, not in our lifetime, and not in the lifetime of the whole universe.

We can conclude that for $N \approx 10^{23}$, the SMI per particle will change monotonically from 0 to 1, and once it reaches the value of 1, it will stay there "forever." This "forever" is not an absolute *forever*, but a practical *forever*. Thus, the concept of irreversibility that we observe is merely an illusion. It is a result of our relatively short lifetime.

These results are sufficient in removing the apparent conflict between the "reversibility" of the equations of motion, and the "irreversibility" of everyday phenomena which we observe in the macroscopic world. We shall further discuss the origin of this apparent conflict in Chapter 6.

We note here that in all of these simulations we did not mention entropy. The common mistake, that has its origin in both Clausius and Boltzmann, is to confuse the behavior of the SMI with entropy. The common mistake has its roots in both Clausius' and Boltzmann's statements, which confuses the behavior of the SMI with entropy. The values of the SMI depends on the distribution (of locations and momenta) at each point of time. The entropy of the system is related to the *maximum* value of the SMI, which is time independent.

4.2 Very large number of particles

Let us proceed with larger N. Figure 4.11 shows the probabilities $P_N(n)$ for larger number of particles. It is seen that the maximum value of $P_N(n)$ *decreases,* as N *increases*.

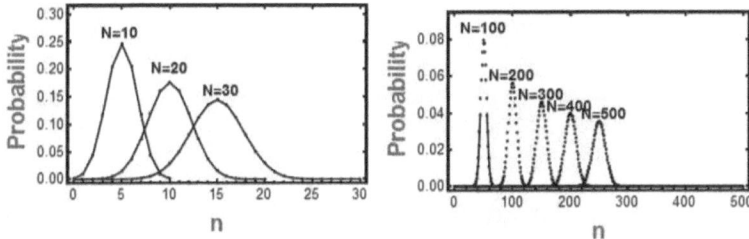

Figure 4.11. The probability of observing n particles in one compartment, and N-n in the other for different N.

It can be shown that the maximal probability *decreases* as $N^{-1/2}$. In practice, we know that when the system reaches the state of equilibrium, it stays there *forever*. The reason is that the *macroscopic state of equilibrium* is not the same as the state for which $n^* = \frac{N}{2}$, but it is this state along with a small neighborhood of n^*, say $n^* - \varepsilon N \leq n \leq n^* + \varepsilon N$, where ε is a small number. For $N = 100$ and $\varepsilon = 0.01$, the probability of finding n in the neighborhood of n^* is about 0.235. For $N = 10^{10}$ particles, we can allow deviations of 0.001% of N and the probability of staying in this neighborhood is nearly one. For more details, see Ben-Naim (2008, 2012).

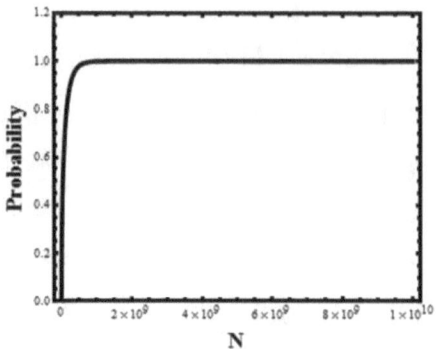

Figure 4.12. The probability of finding n particle in the neighborhood of $n^* = N/2$ in one compartment as a function of N.

In Figure 4.12, we show the probability of finding n between $n^* - \varepsilon N \leq n \leq n^* + \varepsilon N$ as a function of N, with $\varepsilon = 0.0001$. Plotting the probability $P_N(n^* - \varepsilon N \leq n \leq n^* + \varepsilon N)$ as a function of N shows that this probability tends to *one* as N increases. When N is on the order of 10^{23}, we can allow deviations of $\pm 0.00001\%$ of N, or even smaller, yet the probability of finding n at or near n^* will be almost one. It is for this reason that when the system reaches n^* or near n^*, it will stay in the vicinity of n^* most of the time. For N on the order of 10^{23}, "most of the time" means practically, *always*. This is the essence of the Second Law!

The abovementioned specific example provides an explanation to the fact that the system will "always" evolve in one direction, and "always" stays at the equilibrium state once that state is reached. The tendency towards a state of larger probability is equivalent to the statement that events that are supposed to occur more frequently, will occur more frequently. This is plain common sense. The fact that we do not observe deviations from either the monotonic climbing of n towards n^*, or staying close to n^*, is a result of our inability to detect small changes in n- (or equivalently small changes in the SMI, see below). Note that in this section we did not say anything about the entropy changes. Before turning to calculate the entropy changes we repeat the main conclusion of this section. For each N the probability of finding a distribution of particles; $(n, N - n)$ in the two compartments L and R has a maximum at $n^* = \frac{N}{2}$. For very large number of particles the probability of obtaining the *exact* value of $n^* = \frac{N}{2}$ is not very large. However, the probability of finding the system at a small vicinity of $n^* = \frac{N}{2}$ is almost one!

When we say that the system has reached an equilibrium state we mean that we do not *see* any changes that occur in the system. In this example, we mean changes in the *density* of the particles in the entire system. In other experiments when there is heat exchange between two bodies we characterize the equilibrium state as the one in which the temperature is uniform throughout the system and does not change with time.

At equilibrium, the macroscopic density we measure at each point in the system is constant. In the particular system we discussed above the measurable density of the particles in the two compartments is $\rho^* \cong N/2V$. Note that fluctuations always occur. Small fluctuations occur very frequently, but they are so small that we cannot measure them. On the other hand, fluctuations that could be measured are extremely infrequent, and we can practically say that they never occur. This conclusion is valid for very large N.

4.3   The evolution of the SMI in the expansion process and the corresponding change in entropy

Next, we will discuss the relationship between the probabilities of the macrostates and the formulation of the Second Law in terms of the entropy. We rewrite the essential quantities of the examples discussed above in a slightly different way. Instead of n and $N - n$, we define the fractions $p = \frac{n}{N}$, $q = (1 - p) = \frac{N-n}{N}$. p is the fraction of particles in the L compartment and $q = (1 - p)$ is the fraction in the R compartment. We can also think of an ensemble of systems, all having the same macroscopic description in terms of $E, 2V, N$, but the ensemble is prepared in any arbitrary value of p (and hence, $q = 1 - p$). Clearly, the pair of numbers $(p, 1 - p)$ is a probability distribution.

The probability of finding such a distribution of particles $(n, N - n)$ is:

$$\Pr(n, N - n) = \frac{N!}{n!(N-n)!} \left(\frac{1}{2}\right)^N \qquad (4.3)$$

For large n and $(N - n)$ we can take the Stirling approximation:

$$\ln(n!) \approx n\ln n - n$$

and find:

$$\ln\left[\frac{N!}{n!(N-n)!}\right] \approx N\ln N - n\ln n - (N - n)\ln(N - n)$$

$$= N\left[\frac{-n}{N}\ln\frac{n}{N} - \frac{(N-n)}{N}\ln\frac{N-n}{N}\right]$$

$$= -N[p\ln p + (1 - p)\ln(1 - p)] \qquad (4.4)$$

Converting the logarithm to the base 2 by $\ln x = \ln 2 \log_2 x$, we can rewrite eq. (4.3) as:

$$\log_2 \Pr(p) \approx N \text{ SMI}(p) - N \tag{4.5}$$

or equivalently

$$\Pr(p) \approx \left(\frac{1}{2}\right)^N 2^{N \text{ SMI}(p)} \tag{4.6}$$

For the ratio of the probabilies in the initial state $(p = 1)$, and the final state $p = \frac{1}{2}$, we have:

$$\frac{\Pr(final)}{\Pr(initial)} \approx 2^{N \text{ SMI}(final)} \tag{4.7}$$

Note that in the initial state, all particles in L, the SMI($initial$) = 0. Since the SMI($final$) is of the order of one (in this specific example) the probability ratio in (4.7) is extremely large.

The entropy difference for the process is obtained from the SMI *difference* by multiplying by $k_B \ln 2$, i.e.

$$\Delta S = S(final) - S(initial)$$

$$= k_B N \ln 2 \left[SMI(final) - SMI(initial)\right] \tag{4.8}$$

Hence, the entropy of the system at the final state is:

$$S(final) = S(initial) + k_B N \ln 2 \tag{4.9}$$

Note that the $S(initial)$ is the entropy *before* we removed the partition between the two compartments.

From the monotonic relationship between Pr and SMI (either eq. 4.5 or 4.6), it follows that whenever SMI increases, the Pr also increases, and at equilibrium both SMI and Pr attain a maximal value. We have seen that the maximal value of SMI is related to the entropy of the system. Therefore, the answer to the question of *why* the entropy increases (in a spontaneous process in an isolated system), is the same as the answer to the question of why the state of the system evolves towards equilibrium, namely; the probability Pr of the equilibrium state is very large relative to the probability of the initial state.

Note carefully the two "levels" of probabilities. One is the probability distribution of a state described by (p, q). Pr is the probability of finding a state described by (p, q). To distinguish between the two probabilities, I sometimes refer to Pr as a *super probability*. Note also that the answer to the question: "Why does the system evolve towards equilibrium?" is provided by the probability Pr. Because of the monotonic relationship between Pr and the SMI the answer to the question "why does the SMI increase?" is also probabilistic. It is easy to generalize this conclusion for the case of any number of compartments. See Ben-Naim (2008, 2012).

4.4 Summary of facts

Before we move on to the general formulation of the Second Law we summarize what we have found so far from the simple examples of expansion of N particles from volume V to $2V$.

For any N, right after removing the partition we follow the evolution of the system with time. In all the examples, we observed that the particles which were initially confined to one compartment can access the larger volume $2V$. We can ask the following questions:

1. Why do the particles occupy the larger volume?
2. Does the number of particles in the left compartment change monotonically with time?
3. Does the number of particles in the left compartment reach a constant value at equilibrium?
4. How fast does the system reach the equilibrium state?
5. How does the SMI of the system change with time?
6. How does the entropy change with time?

Clearly, the answers to all these questions depend on N. Here are the answers to these questions:

1. The particles will occupy the larger volume $2V$ rather than V because the probability of the states where there are about $N/2$ in each compartment is much larger than the probability of the state where all the particles are in one compartment. This is true for any N. When N is very small there is a relatively large probability that the particles will be found in one compartment. For these cases we cannot claim that the process is irreversible, in the sense that it will never go back to the initial state. For large N, say, of the order 10^6, the probability to return to the initial state becomes so small, that it is practically zero. However, there is always a *finite* probability that the system will visit the initial state. For N of the order of 10^{23}, the probability of visiting the initial state is so small (but still non-zero) that we can safely say that the system will *never* return to the initial state. Never, in the sense of *billions* of *ages* of the universe.

2. The number of particles in L, n, does not change monotonically from N to $N/2$ (or from zero to $N/2$ if we start with all particles in the right compartments). Simulations show, that for large values of N the number n changes *nearly* monotonically towards $N/2$. The larger the N, the more monotonic is the change of n. (For simulated results, see arienbennaim.com, books, Entropy Demystified, simulated games). For N on the order of 10^6 or more you will see nearly perfect, smooth, monotonic change in n.

3. It depends on how one defines the equilibrium state of the system. If we define the equilibrium state when the value of n is equal to $N/2$, then for any n, when n reaches $N/2$ it will not stay there "forever." There will always be fluctuations about the value of $n^* = N/2$. However, one can define the equilibrium state as the state for which n is in a small neighborhood of $n^* = N/2$. In such a definition, we will find that once n reaches this neighborhood, it will stay there for a longer time than in any other state. For N of the order of 10^6 or more, the system will stay in this neighborhood forever. By "forever," we mean many ages of the universe.

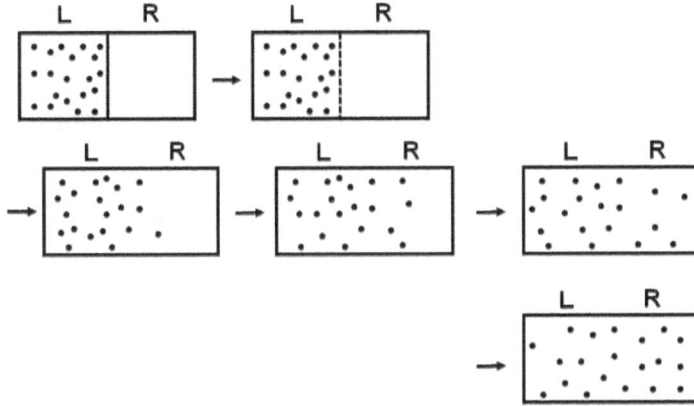

Figure 4.13. The initial, the final, and a few intermediate states in the expansion process.

4. It depends on the temperature and on the size of the aperture we open between the two compartments. In the experiment of Figure 4.13, we remove the partition between the two compartments. However, we could do the same experiment by opening a small window. In such an experiment, the speed of reaching the equilibrium state would depend on the size of the aperture of the window. In any case the principles of thermodynamics do not state anything about the speed of attaining equilibrium.

5. For each distribution of particles $(n, N - n)$ we can define a probability distribution $(p, 1 - p)$, and the corresponding SMI. As the system evolves from the initial to the final state, n will change with time, hence, p will also change with time, hence, the SMI will also change with time. (For simulations, see Ben-Naim, 2010).

For small N, the SMI will start from zero (all particles being in one compartment) and will fluctuate between zero to N bits, i.e. one bit per particle. When N is very large, say 10^6 or more the value of SMI will change nearly monotonically from zero to N bits. There will always be some fluctuations in the value of SMI, but these fluctuations will be relatively smaller for the larger N. Once the system reaches the equilibrium state it will stay there *forever*. Note carefully that the SMI is defined here on the probability distribution $(p, 1 - p)$. For the initial distribution $(1,0)$ the SMI is zero. The SMI defined on the distribution of locations and momenta is not zero, see Chapter 3.

6. The answer to this question is the simplest, yet the most misconstrued one. It is the simplest because entropy is a *state* function, it is defined for a well-defined macroscopic (or thermodynamic) *state* of the system. For the expansion process, the macrostate of the system is defined initially by (E, V, N). The corresponding value of the entropy is $S(E, V, N)$. The final macrostate is characterized by $(E, 2V, N)$, and the corresponding value of the entropy is $S(E, 2V, N)$. In between the two macrostates (E, V, N) and $(E, 2V, N)$ the macrostate of the system is not well-defined. A few, intermediate states are shown in Figure 4.13. While E and N are the same as in the initial state, the "volume" during the expansion process of the gas is not well-defined. It becomes well-defined only when the system reaches an equilibrium state. Therefore, since the volume of the system is not well-defined when the gas expands, the entropy is also not well-defined. We can say that the entropy changes abruptly from $S(E, V, N)$ to $S(E, 2V, N)$, and that this change occurred at the moment the system reaches a final equilibrium state. This is shown schematically in Figure 4.14a.

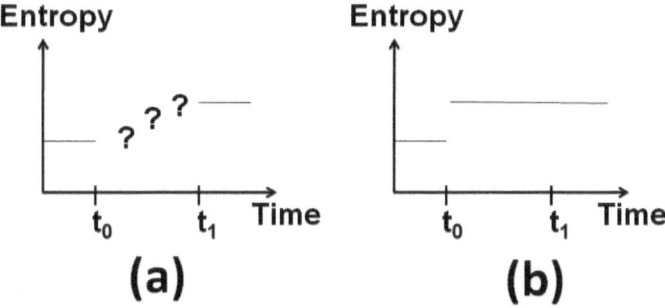

Figure 4.14. Two views of the entropy change after the removal of the partition.

One can also adopt the point of view that when we remove the partition between the two compartments, the volume of the gas changes abruptly from V to $2V$. Although the gas is initially still in one compartment, the total volume *accessible* to all particles is $2V$. If we adopt this view, then at the moment we removed the partition the volume changes from V to $2V$, and the corresponding change in entropy is $S(E, 2V, N) - S(E, V, N)$. This change occurs abruptly at the moment we remove the partition, see Figure 4.14b. Personally, I prefer the first point of view. Initially, it has the value $S(E, V, N)$ *before* the removal of the partition, and it reaches the value of $S(E, 2V, N)$ when the system reaches the new, final equilibrium state. In all the intermediate states the entropy is not defined. Note however, that the SMI is defined for any intermediate states between the initial and the final states. However, the entropy is the maximal value of the SMI (multiplied by the Boltzmann constant and change of the base of the logarithm), reached at the new equilibrium state.

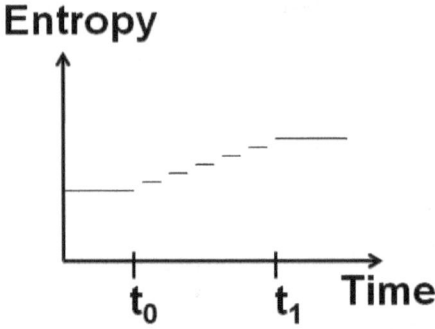

Figure 4.15. The entropy change in a quasi-static process

It should be noted however that we could devise another expansion (referred to as quasi-static) process by gradually moving the partition between the two compartments. In this process the system proceeds through a series of equilibrium states, and therefore the entropy is well-defined at each of the points along the path leading from (E, V, N) to $(E, 2V, N)$. In this process, the entropy of the gas will gradually change from $S(E, V, N)$ to $S(E, 2V, N)$, Figure 4.15. The length of time it takes to proceed from the initial to the final state depends on how fast, or how slow we carry out the process.

Note that the sequences of states in the spontaneous process are different from the quasi-static process. In the latter, the states as well as the entropy of the gas are well defined along the entire path from the initial to the final equilibrium states, whereas in the spontaneous expansion neither the *states*, nor the entropy are defined along the path leading from the initial to the final state.

It is clear by now that the entropy of the system is *never* a function of time. Therefore, all the statements regarding the changes of entropy with time, or the association of entropy with time's arrow are totally unfounded. This statement also includes the behavior of the Boltzmann H-function which is further discussed in chapter 6.

4.5   The formulation of the Second Law for isolated systems

We start with the most well-known formulation of the Second Law based on entropy.

The entropy formulation of the Second Law applies only to isolated systems. We shall formulate it for a one-component system having N particles. If there are k components, then N is reinterpreted as a vector comprising the numbers $(N_1, N_2, ..., N_k)$ where N_i is the number of particles of species i.

For any (unconstrained) isolated system (E, V, N), at equilibrium, the entropy is larger than for any possible constrained equilibrium states of the same system.

Note that this formulation uses only macroscopic quantities. Also, it applies only to equilibrium states. The entropy formulation means that if we remove any of the constraints in any of the initial systems, the entropy will either increase or remain unchanged.

Therefore, an equivalent formulation of the Second Law is:

Removing any constraint from a constrained equilibrium state of an isolated system will result in an increase of the entropy.

Note carefully that we have *defined* entropy for equilibrium systems. The maximum entropy is also a maximum with respect to all *constrained equilibrium states*. This is very different from the maximum of the SMI defined in Chapter 3.

Recall that the SMI may be defined for any distribution (coin, die, particles in a box, etc.). Also, for an (E, V, N) system of particles, we can define the SMI for *any N* and *any* distribution. We look for the particular distribution which maximizes the SMI. The value of the maximum SMI is proportional to the entropy of the system.

To every constrained equilibrium states, corresponds a distribution of locations and momenta of all the particles. The converse of this sentence is not true. Not to every distribution of locations and momenta, corresponds a constrained equilibrium state. An example: Take a system of N particles in a box of length L all moving back and forth at constant speed v, Figure 4.16. For this system the probability distribution of locations and momenta is well-defined. Therefore, the SMI is also well-defined. If the collisions with the walls are perfectly elastic, the state of this system will not change with time. However, this system *is not* an equilibrium state in a thermodynamic sense. Hence, the entropy is not defined for such a system.

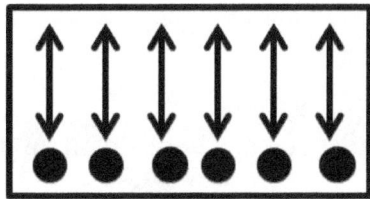

Figure 4.16.   Particles moving up and down at constant velocity

Because of its fundamental importance we repeat the steps leading from the SMI to entropy. The SMI is defined for any system of particles, and for any distribution of locations and momenta. There is *one distribution* that maximizes the SMI. The *value* of the maximum SMI up to a multiplicative constant $(k_B \ln 2)$ is the entropy of the system. The entropy of the system pertains to the equilibrium state and as such it is *not* a function of time and it does not change with time. Note that the entropy is obtained for the *distribution* which maximizes the SMI. The connection with the equilibrium distribution follows from the following considerations. The distribution which maximizes the SMI is the same distribution that maximizes the probability (Pr). We know that any system tends to equilibrium. Therefore, we identify the distribution which maximizes Pr as the *equilibrium* distribution.

Finally, we ask what drives the process when we remove a constraint. The answer; the kinetic energy of the particles. Without the kinetic energy of the particles no process will occur, and the system will not evolve towards a new equilibrium state. The *direction* of the change from one equilibrium state to another is determined by the probability.

In this view, it is clear that the entropy change is not the *cause* of the change but rather a *result* of the change.

In many textbooks, as well as in popular science books, you might find "formulations" of the Second Law as:

> ***The entropy has a tendency to increase.***
>
> ***The entropy of the universe always increases.***
>
> ***The entropy of an isolated system increases until it reaches a maximum.***

All of the above sound similar, but in fact, they are different and all are wrong: The entropy, by itself, does not have a tendency to increase! The entropy of the universe is not defined! The entropy of an isolated system is defined for an equilibrium state. As such, it does not "increase until it reaches a maximum!"

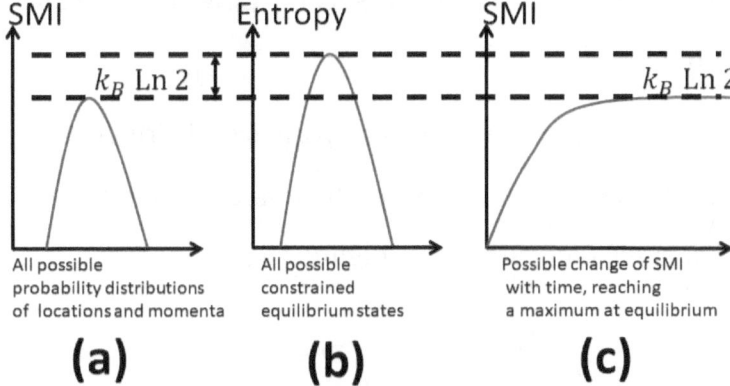

(a)   **(b)**   **(c)**

Figure 4.17. (a) The SMI as a maximum over all possible probability distributions of locations and momenta (b) The entropy as a maximum over all possible constrained equilibrium states. (c) The SMI as a function of time, reaching a maximum at equilibrium. Multiplying the maximum SMI by a constant, we get the entropy of the system.

In Figure 4.17, we schematically show the relationship between the maximum of SMI over all possible distributions, and the maximum of entropy over all entropies of constrained equilibrium states. The maximum SMI corresponds to the equilibrium distribution. The same equilibrium distribution that maximizes the SMI, also maximizes the probability (Pr). The maximum value of the SMI as a function of time is related to the entropy of the system. The connection between the SMI and Pr is by a monotonic function, hence the maximum SMI corresponds to the maximum of Pr. I urge the reader to examine carefully the message transmitted by this figure. It contains the essence of the meaning of entropy, its relationship with the SMI, and the reason for the misinterpretation of the Second Law in terms of entropy.

4.5.1 The entropy formulation of the Second Law for the general case of a c-compartment isolated system

We start with N simple, non-interacting particles distributed in c compartments, of total volume V, such that there are $n_i^{(in)}$ is the initial number of particles in the compartment i, Figure 4.18a.

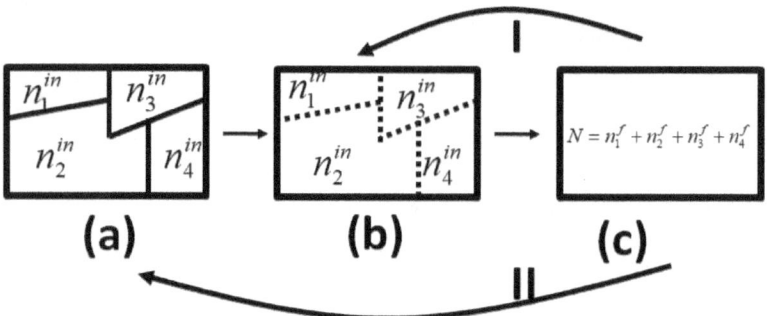

Figure 4.18. (a) A constrained equilibrium system.
(b) The same system as in (a), but with the constraint removed.
(c) The final unconstrained equilibrium state.
The system at state (c) can go back to (b), arrow I, but can never go back to (a), arrow II.

The total volume of the system, the total number of particles, and the total energy of the system are fixed. For this system we have the two equalities:

$$V = \sum_{i=1}^{c} V_i \tag{4.10}$$

$$N = \sum_{i=1}^{c} n_i^{(in)} \tag{4.11}$$

We also assume that each compartment can accommodate any number of particles $0 \le n_i \le N$. The vector $\boldsymbol{n}^{(in)} = \left(n_1^{(in)}, n_2^{(in)}, \ldots, n_c^{(in)} \right)$ may be referred to as a *partition* of the number N into c numbers $n_i^{(in)}$ such that the sum of $n_i^{(in)}$ is equal to N. In this section, we use the word *partitions* for the barriers separating between the compartments, and the vector $\boldsymbol{n}^{(in)}$ will be referred to as the initial state distribution of the system.

Initially, we have a *constrained equilibrium state*. The entropy of the system as a whole is given by:

$$S_{Total}^{(in)} = \sum_{i=1}^{c} S_i^{(in)} \tag{4.12}$$

where $S_i^{(in)}$ is the entropy of the compartment i.

We now show that whenever we remove all the partitions between the compartments the total entropy of the system will increase. This will be the basis on which we shall establish the entropy formulation of the Second Law for the isolated system.

The process is shown in Figure 4.18. Initially, we have a constrained equilibrium state (before the removal of the partitions). Immediately after the removal of the partitions, the system is shown in Figure 4.18b. At this point the system is not in an equilibrium state. It is still characterized by the vector ($n_1^{(in)}, n_2^{(in)}, ..., n_c^{(in)}$). After some time, the system will reach a new (unconstrained) equilibrium state Figure 4.18c. We denote by $(n_1^{(f)}, n_2^{(f)}, ..., n_c^{(f)})$ the final equilibrium state. It should be noted that $n_i^{(f)}$ is the final number of particles in the compartment i, in the *absence* of the partitions (shown as dashed lines in figure 4.18).

We already know that the locational SMI for each particle is maximum for the uniform distribution, i.e. the probability of finding a single particle in any element of volume dV at a point $\mathbf{R}$, is:

$$f(\mathbf{R})dV = \frac{dV}{V} \tag{4.13}$$

$f(\mathbf{R})$ is independent of the locational vector $\mathbf{R}$.

If we choose dV to be small enough so that at most one center of a particle can be accommodated in this volume, the probability of finding any particle in dV is NdV/V. We denoted by $\rho = \frac{N}{V}$ the density of particles at the final equilibrium state. The average number of particles in each compartment of volume V_i (in the absence of the partitions) is thus:

$$n_i^{(f)} = \rho V_i = \frac{NV_i}{V} \tag{4.14}$$

Thus, when we remove the partitions separating all the compartments, the system will move to a new equilibrium state. At this equilibrium state, the uniform locational distribution has the largest probability. Uniform distribution implies that the average number of particles in any compartment is proportional to the volume of that compartment, i.e. $n_i^{(f)} = \rho V_i$ or $x_i^{(f)} = \frac{V_i}{V}$, where $x_i^{(f)}$ is the molar fraction of particles in compartment i. The corresponding change in entropy can be calculated from the entropy function, see Chapter 3. The difference in entropy in the process in Figure 4.18, is:

$$\Delta S = S(E, V, N) - \sum_{i=1}^{c} S\left(E_i, V_i, n_i^{(in)}\right)$$

$$= N k_B \ln \left(\frac{V}{N} \alpha \right) - \Sigma N_i k_B \ln \left(\alpha \frac{V_i}{n_i^{(in)}} \right) \qquad (4.15)$$

In eq. (4.15) we included in α all the factors which do not change in the process, in particular $E_i / n_i^{(in)}$ which is proportional to the temperature T. Denoting the *mole fractions* as:

$$x_i^{(in)} = \frac{n_i^{(in)}}{N} \quad , \quad x_i^{(f)} = \frac{n_i^{(f)}}{N} \qquad (4.16)$$

We can rewrite eq. (4.15) as:

$$\Delta S = N k_B \sum_{i=1}^{c} x_i^{(in)} \ln \left[\frac{V n_i^{(in)}}{N V_i} \right]$$

$$= N k_B \Sigma_{i=1}^{c} x_i^{(in)} \ln \left[x_i^{(in)} / x_i^{(f)} \right] \geq 0 \qquad (4.17)$$

The equality holds if, and only if, $x_i^{(in)} = x_i^{(f)}$ for all i. The inequality in (4.17) follows from the fact that apart from the Boltzmann constant and the change in the base of the logarithm, this is the Kullback-Leibler [see Ben-Naim(2017b)] distance between the two distributions $\boldsymbol{x}^{(in)}$ and $\boldsymbol{x}^{(f)}$.

We conclude this section by answering the two questions:

1. Why did a spontaneous process occur?

2. Why does the entropy increase?

The answer to the first question is straightforward. Accepting the relative frequency interpretation of probability, a system will always spend more time in states having higher probability. When the system is macroscopic (i.e. N of the order of Avogadro number), the probability of the final equilibrium state is overwhelmingly larger than the probability of any other state, therefore the system will move with almost certainty to a new equilibrium state.

Regarding the second question, first we should note that the entropy formulation of the Second Law is valid for an isolated system. Because of the monotonic relationship between the probability and the SMI of the distribution, whenever the probability ratio is larger than one, the difference in the SMI between the initial and the final states will be positive. Therefore, the difference in the entropy is also positive.

We emphasize again that while the first answer is valid for any spontaneous process in any thermodynamic system, the formulation of the Second Law in terms of the entropy is valid only for isolated systems.

Let us repeat a similar experiment as follows: Suppose we start with a system having fixed values of E, V, N. We divide it into c compartments of equal volumes, V/c each having different number of particles n_i, such that $\sum n_i = N$. The initial entropy of each compartment is denoted by $S_i^{(in)}$ and the total entropy of the system, assuming that each compartment is macroscopic and at equilibrium, is

$$S^{(in)}(E, V, N; \boldsymbol{x}^{(in)}) = \sum_{i=1}^{c} S_i^{(in)} \qquad (4.18)$$

where $\boldsymbol{x}^{(in)}$ is the distribution defined by its components $x_i^{(in)}$.

When we remove all the partitions between the compartments, keeping E, V, N constant, the system will evolve into a new equilibrium state with an entropy value which is *larger* than the initial entropy, and for which the distribution of particles is uniform, i.e. $x_i^{(f)}$ and

$$S^{(f)}\left(E,V,N;x^{(f)}\right) \geq S^{(in)}\left(E,V,N;x^{(in)}\right) \tag{4.19}$$

Again, we emphasize that the reason for the evolution of the system, from the initial to the final state:

$$x^{(in)} \to x^{(f)} \tag{4.20}$$

is probabilistic, and the change in entropy is always positive (under the condition that the system is isolated).

Note that in this example, starting from *any* initial distribution $x^{(in)}$ the system will always evolve towards the *uniform distribution* $x^{(f)}$. The reason is that the uniform distribution is the one that maximizes the probability $\Pr(x)$, or equivalently, the corresponding SMI. A more general formulation is discussed in Ben-Naim(2017b).

We can now state the probability formulation of the Second Law as follows: We first write the relationship between the entropy difference in the process, and the probability ratio:

$$\frac{\Pr(final)}{\Pr(initial)} = \exp\left[[S(final) - S(initial)]/k_{\mathrm{B}}\right] \tag{4.21}$$

Thus, in an isolated system, since $\Delta S = S(final) - S(initial) > 0$, the probability ratio is very large, of the order of $\mathrm{Exp}[N]$. As we shall soon see the probability formulation is much more general than the entropy formulation. Note that the state *"final"* on the two sides of eq. (4.21) is the final *equilibrium* state. On the other hand, the state *"initial"* on the right-hand side of the equation is the state *before* the removal of the constraints, whereas the state *"initial"* on the left- hand side is the state right *after* the removal of the constraints.

4.6 The formulation of the Second Law for a *T,V,N* System

In the previous section, we discussed the entropy formulation of the Second Law, and emphasized that it applies to a well-defined *isolated* system denoted by (E, V, N).

A more convenient system to work with is a closed system having a fixed *volume V*, and constant *temperature T*. Such a system can exchange heat with its surroundings. Normally, we keep the temperature of the *system* fixed by placing it in a heated *bath*. The heat bath is supposed to be very large compared with the system, such that when there is a small exchange of heat between the system and the bath, the bath's temperature is not affected.

Before we formulate the Second Law for such a system we define the Helmholtz energy by:

$$A = E - TS. \tag{4.22}$$

Here we have on the right-hand side of the equation the energy E, the temperature T and the entropy S. Since S is *only* defined for equilibrium systems, so is the Helmholtz energy.

This definition of the Helmholtz energy is valid for any system at equilibrium. In this definition we did not specify the independent variables with which we characterize the system. We have the liberty to choose any set of independent variables, say $(E, V, N), (T, V, N)$ or (T, P, N). However, if we want to formulate the Second Law in terms of the Helmholtz energy we must choose the independent variables (T, V, N). For a (T, V, N) system the Helmholtz energy formulation of the Second Law is:

For any (T, V, N) system at equilibrium the Helmholtz energy has a minimum over all possible constrained equilibrium states of the same system.

Recall that the entropy formulation of the Second Law was valid only for *isolated* systems, i.e. for (E, V, N) systems. The Helmholtz energy formulation is valid only for systems that are isothermal (constant T), and isochoric (constant V), as well as closed (constant N).

An equivalent statement of the Helmholtz energy formulation is:

Removing any constraint from a constrained equilibrium state in a (T, V, N) system will result in a decrease in Helmholtz energy.

The connection between the Helmholtz energy change and the probability ratio is:

$$\frac{\Pr(final)}{\Pr(initial)} = \exp\left[-[A(final) - A(initial)]/k_{\mathrm{B}}T\right] \qquad (4.23)$$

Since $[A(final) - A(initial)] < 0$, and large, the probability ratio is very large. The derivation of this formulation is discussed in details in Ben-Naim (2017c).

4.7 The formulation of the Second Law for a *T,P,N* system

Next, we briefly state Gibbs energy formulation of the Second Law. This formulation is valid for the processes carried out at constant temperature (T), and pressure (P). The system is still closed, i.e. N (or $N_1, ..., N_c$) is constant.

Before we formulate the relevant Second Law for such a system, we define the Gibbs energy by:

$$G = E - TS + PV \qquad (4.24)$$

As in section 4.6 we note here that the definition of the Gibbs energy applies to any thermodynamic system at equilibrium. We have the liberty to choose the independent variables characterizing the system. However, for the Gibbs energy formulation of the Second Law we must choose the specific independent variables (T, P, N). Here is the Gibbs energy formulation for the T, P, N system:

For any (T, P, N) system at equilibrium the Gibbs energy has a minimum over all the possible constrained equilibrium states of the same system.

It is important to emphasize that this formulation of the Second Law is valid for a system at constant temperature (isothermal), constant pressure (isobaric), and closed (i.e. impermeable to particles).

An equivalent statement of the Gibbs energy formulation is:

Removing any constraint from a constrained equilibrium state of a (T, P, N) system will result in a decrease in the Gibbs energy.

As we emphasized in section 4.6, the Gibbs energy is defined for a well-defined thermodynamic system at equilibrium. It is not true that the Gibbs energy decreases in every process occurring in any system.

Again, we note that underlying this formulation of the Second Law is a more fundamental principle. As in eq. (4.23), the ratio of the probabilities is now related to the difference in the Gibbs energies, i.e

$$\frac{\Pr(final)}{\Pr(initial)} = \exp\left[-[G(final) - G(initial)]/k_{\mathrm{B}}T\right] \qquad (4.25)$$

where $\Delta G = G(final) - G(initial) < 0$. The same comment we made following eq. (4.23) also applies in this case.

4.8. Summary of Chapter 4

In this chapter, we saw that the main "driving force" for any spontaneous process in a thermodynamic system may be attributed to probability. The ratio of the probabilities in the initial and final states is related to the entropy change in an isolated system, to the Helmholtz energy change in a T, V, N system, and to the Gibbs energy change in a T, P, N system.

We present here the main thee equations (3.21), (3.23) and (3.25):

$$\frac{\Pr(final)}{\Pr(initial)} = \exp\left[[S(final) - S(initial)]/k_{\mathrm{B}}\right]$$

$$\frac{\Pr(final)}{\Pr(initial)} = \exp\left[-[A(final) - A(initial)]/k_{\mathrm{B}}T\right]$$

$$\frac{\Pr(final)}{\Pr(initial)} = \exp\left[-[G(final) - G(initial)]/k_{\mathrm{B}}T\right]$$

The first is valid for an (E, V, N) system, the second is valid for a (T, V, N) system, and the third for a (T, P, N) system. The first equation reduces to Boltzmann's formulation when all the microscopic states have equal probabilities, in which case the probability ratio is equal to the ratio: *W(final)/W(initial)*. Note that sometimes, *W* itself is equated to the probability (Pr), and not the ratio. This is not true since the probability is a number between zero and one, whereas *W* could be any number.

It is clear that the probability formulation of the Second Law is far more general than any of the thermodynamic formulations in terms of either entropy, Helmholtz energy or Gibbs energy. It is also clear that the probability formulation is not in conflict with the time reversibility of the equations of motion. One important advantage of the probability formulation is that the Second Law does not apply for any system which has a "free will." This conclusion debunks claims by many authors, e.g. Atkins (2007) that entropy or the Second Law "controls" or "drives" our thoughts, feelings and creation of arts. The fact is that no one has ever shown that either entropy or the Second Law has anything to do with "thinking, feelings or creation of arts." For details see Ben-Naim (2017c).

Chapter 5. A few simple examples

In this chapter we discuss three simple examples that are discussed in almost any book on thermodynamics. In fact, these three examples were used in the very formulations of the Second Law. The processes we discuss are the expansion of ideal gas, the mixing of two different gases, and the heat transfer from a hotter to a colder body. These processes are carried out in an isolated system. In each of these processes we will calculate the entropy change, the SMI change, and examine the question of reversibility of the process, and whether or not the entropy can be said to be changing with time.

5.1   The simplest expansion of an ideal gas from V to $2V$

The expansion of ideal gas was already discussed in Section 4.3. For the expansion from V to $2V$, the change in entropy is:

$$\Delta S = N\, k_B \ln 2 \qquad (5.1)$$

The probability ratio is:

$$\frac{\Pr(final)}{\Pr(initial)} \approx \exp[N] \qquad (5.2)$$

This is a huge number.

A few comments on this process are now in order.

First, note that the entropy change in the process $V \rightarrow 2V$ in an isolated system is *independent* of temperature. This is important. Some people who interpret entropy as a measure of the "spreading of energy" would conclude that if you have the same expansion process, but at two different temperatures, say $T_1 = 300K$, and $T_2 = 400K$, the spread of energy in the former is less than in the latter. Therefore, one is led to conclude that the change in entropy at the higher temperature should be *larger* than that of the lower temperature. This conclusion is not correct, either experimentally or theoretically.[16]

The second comment concerns the independence of ΔS in this process on the *kind* of gas. Again, if one interprets entropy as spreading of energy one might conclude that particles with larger mass have more kinetic energy than particles having a smaller mass.

Note however that when the same expansion process occurs with non-ideal gases the entropy change will depend on the kind of particles. The reason is that with a non-ideal gas the entropy change due to the *locational* change in the SMI is independent of the kind of particles. However, there is another contribution to the change of entropy in this process. The average interaction energy between the particles changes when the gas expands from V to $2V$. This part of the change in entropy will depend on the type of the particles in the gas.

I raised these comments here although they are quite trivial. They are far from trivial when we discuss the entropy change of two different ideal gases. I will add here that the "informational" interpretation of the change of entropy in the expansion process, eq. (5.1) is quite simple. In this process there is a loss of one bit of information for each particle, hence, N bits for N particles. We shall see that exactly the same interpretation applies for the mixing of ideal gases.

5.2   Entropy of mixing of two ideal gases

The mixing of two different ideal gases as illustrated in Figure 5.1a, is another example in which the entropy change can easily be calculated. It is also a process which is misunderstood by most authors who write about it including Gibbs, who was the first to thoroughly study this process (see below).

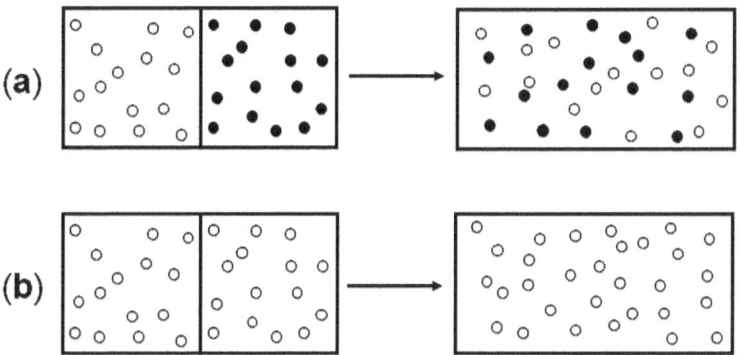

Figure 5.1.   (a) Mixing of two different kinds of molecules
(b) "Mixing" the same kind of molecules

If we have one mole of A in a volume V, at a temperature T, and another mole of B in a volume V, and temperature T, then the entropy change for the process shown in Figure 5.1a is:

$$\Delta S = 2R\ln 2 \tag{5.3}$$

While there is no question about the *value* of ΔS for this process, there is great confusion regarding its interpretation. Here are the three main interpretations of this increase in the entropy.

(i) The experimental interpretation

Mixing is viewed as a process of disordering. We all agree that the mixture of A and B on the right-hand side of Figure 5.1a, is a more disordered state than the system on the left-hand side of the Figute, where the two components are separated.

Most textbooks interpret entropy as a measure of disorder. Positive change in entropy means that the system became more disordered. Therefore, one concludes that the entropy change in the mixing process is *due* to the mixing, i.e. to increasing disorder, hence, the term "entropy of mixing." This interpretation sounds quite reasonable, nevertheless, it is wrong! Gibbs was puzzled by the fact that the "entropy of mixing" is independent of the types of the two gases as long as they are different. See below for more details and Ben-Naim (2017a, 2017c).

(ii) *The statistical mechanical interpretation*

Using the methods of statistical mechanics of ideal gases, it is easy to calculate the entropy change for the process as shown in Figure 5.1a (for details, see Ben-Naim, 2008). The result is: $\Delta S = 2R\ln \frac{2V}{V} = 2R\ln 2$.

This is exactly the same result we obtain from thermodynamics. However, in this calculation the number 2 under the logarithm is a result of the increase in the *accessible volume* for each of the $2N$ particles in the system, not a result of the *mixing*.

(iii) *The information interpretation*

This is essentially the same as the statistical mechanical interpretation. However, now we calculate the difference in the SMI for the two states of the system. If we use the logarithm to the base 2 we get; change of SMI $=$ $2N\log_2 2 = 2N$ bits. This is exactly the same interpretation given to the expansion process. Here we have $2\ N$ instead of N particles.

Gibbs, who was the first to analyze the so-called "entropy of mixing" was puzzled by the fact that the entropy of mixing is *independent* of the kind of molecules, for as long as A and B are distinguishable molecules. Note that this conclusion is valid for ideal gases. If you mix two liquids, two solids, or two non-ideal gases you will find that the entropy change in the mixing process, *depends* on the kinds of molecules A and B. In the latter cases the entropy of mixing is a result of two factors: One is the change in the volume accessible to each particle. The second is due to the change in the interaction energies for the pairs A-A, A-B, and B-B.

When we mix two *ideal gases*, we completely neglect all intermolecular interactions. In this case the entropy change, in the mixing process shown in Figure 5.1a is due only to the change in the accessible volume. Hence, the so-called "entropy of mixing" of ideal gases is nothing but the entropy of *expansion*. Each gas expands from V to $2V$. In this view there is nothing puzzling in the fact that the "entropy of mixing" of ideal gases is independent of the kind of molecules A and B. It is unfortunate that Gibbs himself failed to see that *mixing* of two different ideal gases is nothing but expansion of each gas from V to $2V$, and that the mixing, by itself, does not contribute anything to the thermodynamics of mixing (of ideal gases). Mixing, as well as demixing of ideal gases can occur with positive, negative, or zero change in entropy, Figure 5.2.

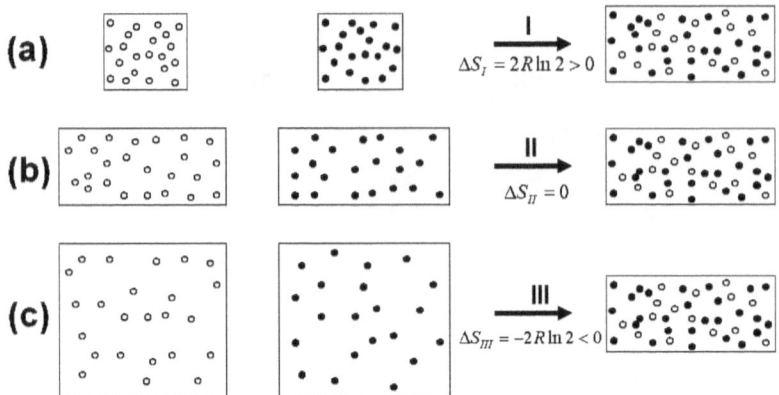

Figure 5.2. Three processes of mixing with positive, zero, and negative changes in entropy

There is another process for which Gibbs reached a wrong conclusion. This is the process shown in Figure 5.1b. This process is exactly the same as in 5.1a, except that in this case the two compartments contain the same molecules, say A. In this process, there is no change in entropy when we remove the partition between the two compartments. This result ($\Delta S = 0$) is accepted by everyone. There is however disagreement regarding the interpretation of this result. Here are the three interpretations of the process shown in Figure 5.1b.

(i) The thermodynamic interpretation

Here, the result $\Delta S = 0$ is straightforward, if not obvious. After removing the partition in Figure 5.1b, we do not observe any process. If we do not *see* anything happening, then it is natural to conclude that there is *no process*. If there is no process, then there should be no change in entropy: An obvious conclusion. Obvious indeed, but a wrong conclusion!

(ii) The statistical mechanical interpretation

If we write the partition function for the initial and the final states, we can calculate the change in the Helmholtz energy for this process, and from its temperature derivative, we can calculate the entropy change [For details, see Ben-Naim (2008)]. The result is:

$$\Delta S = 2Nk_B\ln2 + k_B\ln\frac{(N!)^2}{(2N)!}.$$ (5.4)

One can easily prove that this entropy change must always be positive, see Ben-Naim (2008).

Thus, statistical mechanics tells us that there are *two processes* going on in the seemingly non-process shown in Figure 5.1b. One is expansion from V to $2V$, the second is the change in the number of indistinguishable particles. We started with N indistinguishable particles in the left compartment, and another N indistinguishable particles in the right compartment. After removing the partition between the two compartments we have $2N$ indistinguishable particles. Thus, we have two contributions to the entropy change, one due to expansion and the second, due to change in the number of indistinguishable particles.

But, why is $\Delta S > 0$, in disagreement with the experimental result that $\Delta S = 0$? The answer is that $\Delta S > 0$ for any *finite number* of particles. However, when N is very large, such that we can use the Stirling approximation for $\ln N!$, the two terms cancel each other.

Thus, we see that the two contributions to $\Delta S > 0$ for this process cancels out for large N, resulting in $\Delta S \approx 0$.

(iii) The informational interpretation

This interpretation is essentially the same as the statistical interpretation. However, conceptually it is slightly different. As for the mixing process shown in Figure 5.1a we *lose* one bit per particle due to the expansion process. On the other hand, we *gain* one bit per particle due to the change in the number of indistinguishable particles. The net change is zero bits per particle (provided that N is very large).

As we noted above, Gibbs was puzzled by the fact that the entropy of mixing of two different gases is the same, independently of the kind of molecules in the two compartments. As long as the two gases are different, there is a finite change in entropy. When the two gases are the same there is no change in entropy. It should be said that Gibbs understood that the difference in the entropy changes for the two processes in Figure 5.1 is a result of the indistinguishability of the molecules. He did not see any paradox in the fact that ΔS changes discontinuously when we do the experiment with different kinds of particles, or with the same kind of particles. Later, people were puzzled by this discontinuous "jump" in ΔS from the finite value of $2R\ln2$ to zero. This puzzle is sometimes referred to as the *Gibbs paradox*. Unfortunately, there has never been a paradox, and even Gibbs himself did not see it as a paradox. Gibbs understood that atoms and molecules can either be distinguishable or indistinguishable. There is no continuous change from being distinguishable to being indistinguishable. Such a continuous change could be envisaged for two different objects, say two labelled balls. One can imagine that the label is being continuously reduced to zero. This process will transform two *different* macroscopic objects into two *identical* objects. However, such a transformation cannot be achieved in the microscopic world of atoms and molecules. These particles are either distinguishable or indistinguishable. There is no continuous passage from one to another.

Notwithstanding this fundamental aspect of the microscopic world, Gibbs erred in his conclusion regarding the two processes in Figure 5.1. For the *proper* mixing in Figure 5.1a, Gibbs wrote that although it is an irreversible process, it can be reversed. Of course, one needs to invest energy to do so. The process in Figure 5.1a will not be reversed spontaneously.

On the other hand, for the process in Figure 5.1b, Gibbs wrote that its reversal is "entirely impossible." The reason is clear. Once we remove the partition between the two compartments, each particle can wander in the entire volume $2V$. There is no way to "reverse" the process, in the sense of bringing back each particle originating from the left compartment, to the same compartment, the same goes for the particles originating from the right compartment. Why? Because the particles are *indistinguishable*. Once we remove the partition there is no way we can tell which particle originated from which compartment. Hence, Gibbs concluded that this reversal is "entirely impossible."

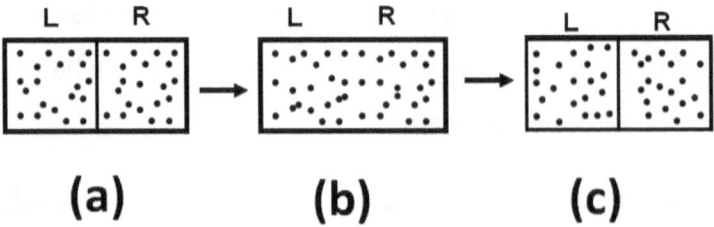

Figure 5.3. The same process as in Figure 5.1b, but now the partition is returned back to its original place

Ironically, Gibbs failed to see that for the *same reason* he claimed that the reversal of the process in Figure 5.1b is "entirely impossible," it is in fact, *trivially* possible. One can effortlessly, and with no investment of energy, reverse this process by simply putting the partition at its original place, Figure 5.3. By doing so, we will get two compartments having (nearly) the same number of particles. No one can claim that the final state is not the same as the original state we started in Figure 5.1b. The reason is simply the *indistinguishability* of the particles.

5.3 Entropy changes associated with change in the intramolecular interactions

Up to this point, we discussed ideal gas systems, i.e. systems with no intermolecular interactions or negligible interactions. We discuss here a particularly simple process in which the intermolecular interaction changes. We assume that the density of the system is very low so that interaction among three or more particles can be neglected, but pair interactions are still operative. One can show (Ben-Naim (2008, 2017c) that the change in SMI associated with turning on the (pairwise) interactions is:

$$\Delta S = -\frac{N(N-1)}{2V^2} \int \int P(\pmb{R}_1, \pmb{R}_2) \ln g(\pmb{R}_1 \pmb{R}_2) d\pmb{R}_1 d\pmb{R}_2 \qquad (5.5)$$

Where $P(\pmb{R}_1, \pmb{R}_2)$ is the pair distribution function and $g(\pmb{R}_1, \pmb{R}_2)$ is the pair correlation function. We now design a process where the SMI changes only due to changes in the interaction energy among the particles. We start with two boxes of equal volume V and equal number of particles N. The particles in the two boxes are different, but their intermolecular potential function $U(R)$ is the same as in Fig 3.12, say two isotopes of argon.

We now bring the two systems into one box of the same volume V. Note that since the particles in the two boxes were different, we will have a *mixture* of N particles of one kind, and N particles of the second kind. We have seen that this mixing process had no effect on the SMI of the system. There is also no change in the volume accessible to each particle in the system. Although the total density of particle has changed in this process we still assume that higher order correlations between particles may be neglected. If the process is carried out at constant temperature, then neither $P(\pmb{R}_1, \pmb{R}_2)$ nor $g(R_1, R_2)$ will change in the process. The change in the SMI in this process is thus:

$$\Delta S = S_f - S_i = \left[-\frac{2N(2N-1)}{2V^2} + \right.$$

$$\left. 2\frac{N(N-1)}{2V^2} \right] \int \int P(\pmb{R}_1, \pmb{R}_2) \ln g(\pmb{R}_1 \pmb{R}_2) d\pmb{R}_1 d\pmb{R}_2 \qquad (5.6)$$

Note that the only thing that has changed is the *total* number of *pairs* of interacting particles; from about N^2 pairs to about $2N^2$ pairs.

It should be noted that in all of the examples discussed in the previous sections, the processes involved ideal gases, i.e. no interactions and no changes in internal degrees of freedom. In such a process, keeping the temperature constant is equivalent to keeping the total energy constant. Therefore, we could calculate the changes in SMI either from the entropy function.

For systems of interacting particles, the situation is quite different. The change in the SMI depends on whether we carry out the process at constant temperature or at constant energy (isolated system). For more details, see Ben-Naim (2008, 2017b).

5.4 A process involving pure change in momentum distribution

In the previous examples we discussed changes in SMI due to the changes in *locational distribution*. We now turn to discuss the simplest process where changes in momentum distribution is involved. This is also important from the historical point of view. This is a classical process for which the Second Law was formulated.

Consider the following process. We start with two systems of ideal gases, each containing N particles in a volume V, but the temperatures are different, say $T_1 = 100K$ and $T_2 = 300K$. The assumption of equal N, and equal volume V is not necessary. Experimentally, we observe that the temperature of the hot gas will get lower, and the temperature of the cold gas will get higher. At equilibrium, we shall have a uniform temperature of $T = 200K$ throughout the system.

Clearly, heat or thermal energy is transferred from the hot to the cold gas. The change in entropy in this process can be calculated either from thermodynamics, or from the entropy function. Initially, the entropy of the system is $S(T_1, V, N) + S(T_2, V, N)$ and in final state the entropy of the system is $2S(T, V, N)$.

The entropy change for this process is:

$$\Delta S = k_B \left[\frac{3}{2}(2N) \ln T - \frac{3}{2} N \ln T_2 - \frac{3}{2} N \ln T_1 \right]$$

$$= k_B \left[\frac{3}{2} N \ln \frac{T}{T_1} + \frac{3}{2} N \ln \frac{T}{T_2} \right] \tag{5.7}$$

where the two terms on the right-hand side correspond to the changes in the entropy of the two systems. Since T is the arithmetic average of T_1 and T_2 we have:

$$\Delta S = k_B \left[\frac{3}{2} N \ln \frac{T_1 + T_2}{2T_1} + \frac{3}{2} N \ln \frac{T_1 + T_2}{2T_2} \right]$$

$$= k_B \frac{3}{2} N \ln \frac{(T_1 + T_2)^2}{4 T_1 T_2} = k_B \frac{3}{2} N \ln \frac{\left(\frac{T_1 + T_2}{2} \right)^2}{T_1 T_2}$$

$$= k_B 3 N \ln \left[\frac{\frac{T_1 + T_2}{2}}{\sqrt{T_1 T_2}} \right] \geq 0 \tag{5.8}$$

The last inequality follows from the inequality about the arithmetic and the geometric average, i.e.

$$\frac{T_1 + T_2}{2} > \sqrt{T_1 T_2} \tag{5.9}$$

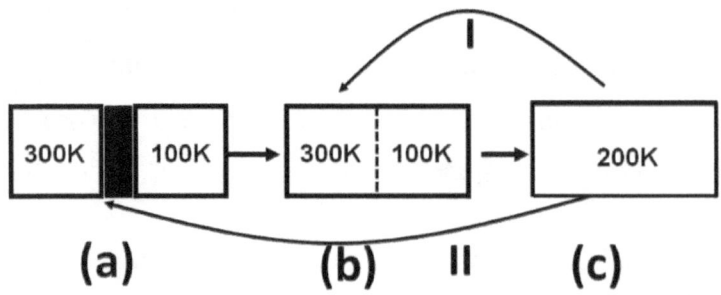

Figure 5.4. The initial, the final, and an intermediate state in the heat transfer process

Thus, in the spontaneous process as depicted in Figure 5.4, the entropy of the entire system increased. It should be noted however that the positive change in entropy as calculated in eq. (5.8) is not an explanation of why the process proceeds spontaneously from the initial state to the final state. In both the initial and final states, the momentum distribution is such that it maximizes the SMI of the system. The Second Law of thermodynamics states that the entropy in the final state must be larger than the entropy in the initial state. To understand why the system proceeds spontaneously from the initial to the final states, we must appeal to probabilistic argument.

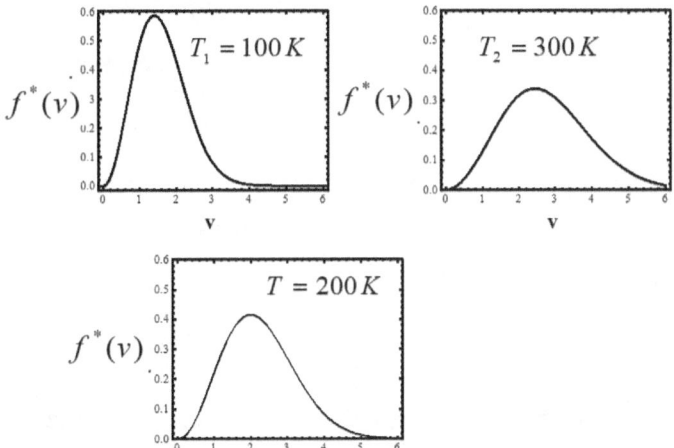

Figure 5.5. The absolute velocity (speed) distribution of the two systems at the initial state (a), and in the final state (b).

Figure 5.5 shows the distribution of the speeds (i.e. the absolute speed $v = \sqrt{v_x^2 + v_y^2 + v_z^2}$), for the two systems at T_1 and T_2, in the initial state. The final distribution (at T) is the distribution that maximizes the SMI. It is also the distribution that maximizes the probability of the velocity distribution. Thus, the passage from the initial to the final state is understood as a result of the overwhelming larger probability of the final state. For more details see Ben-Naim (2008).

Because of the probabilistic argument used in the explanation of the process of heat transfer, we can conclude as we concluded in the previous examples that the system *can* go from the final state c to the initial state just after the two subsystems were brought to thermal equilibrium, Figure 5.4b. This event will be extremely rare; it is not impossible but highly improbable. We can also say that the process is not irreversible in an absolute sense, but only in a practical sense. However, going back to state a in Figure 5.4a is impossible.

Regarding the entropy, once we bring the two subsystems into thermal contact, Figure 5.4b, the entropy of the system becomes undefined. Alternatively, we can say that the states of the system just after the thermal contact are part of the total states of the combined system. Therefore, at the moment we removed the constraint (i.e. allow for the transfer) the entropy attains its final value $2S(T, V, N)$.

Can the value of the entropy return to its initial value of $S(T_1, V, N) + S(T_2, V, N)$? The answer is a No! For this to happen the system has not only to return to its initial state b in Figure 5.4b (which is possible though extremely improbable), but also stay at that state which means that the constraint will be placed as in Figure 5.4a. This will never occur spontaneously.

Chapter 6. Boltzmann's H-theorem; its criticisms, answers and the seeds of an enormous misconception about entropy

In 1877 Boltzmann proved a remarkable theorem known as the H-theorem. He defined a function $H(t)$ and proved that it decreases with time and reaches a minimum at equilibrium.

This behavior of the H-theorem was meant to represent the behavior of the entropy, i.e. the increase of entropy with time, which for over a hundred years was considered to be the ultimate formulation of the Second Law. For instance, Davies (1995) writes:

" *He [Boltzmann] discovered a quantity, defined in terms of the motions of the molecules that provided a measure of the degree of chaos in the gas. This quantity, Boltzmann proved, always increases in magnitude as a result of the molecular collision, suggesting it be identified with thermodynamic entropy. If so, Boltzmann's calculation amounted to a derivation of the second law of thermodynamics from Newton's laws.* "

This is clearly not true. First, the quantity defined by Boltzmann is not a "measure of the degree of chaos." Second, Boltzmann did not prove the Second Law from Newton's laws. This is a typical erroneous statement of Boltzmann's H-theorem which dominates the literature.

Boltzmann's theorem drew serious criticisms, the most important ones being, the "reversal paradox," and the "recurrence paradox."

Boltzmann correctly answered his critics. These answers were correct for as long as the H-function is viewed as a SMI (which of course was not known as such in the late 19th century). Unfortunately, both Boltzmann and his critics erred in interpreting the $-H(t)$ function as entropy. This error still dominates the literatures to this day. In Appendix C we provide a few quotations on this topic from recent literature.

In this chapter we briefly present the H-theorem, the main criticism, and Boltzmann's answer. We will point out where Boltzmann went wrong and why the function $-H(t)$ is not entropy, and why the H-theorem does not represent the Second Law.

6.1   Boltzmann's H-theorem

Boltzmann defined a function $H(t)$ as:

$$H(t) = \int f(v,t) \log[f(v,t)] \, dv \qquad (6.1)$$

and proved a remarkable theorem known as Boltzmann's H-theorem. Boltzmann made the following assumptions:

1. Ignoring the molecular structure of the walls (perfect smooth walls).

2. Spatial homogenous system or uniform locational distribution.

3. Assuming binary collisions, conserving momentum and kinetic energy.

4. No correlations between location and velocity (assumption of molecular chaos).

Details of the assumptions and the proof of the theorem can be found in many textbooks. Basically, Boltzmann proved that:

$$\frac{dH(t)}{dt} \leq 0 \tag{6.2}$$

and at equilibrium, i.e. $t \to \infty$:

$$\frac{dH(t)}{dt} = 0 \tag{6.3}$$

Boltzmann believed that the behavior of the function $-H(t)$ is the same as that of the entropy, i.e. the entropy always increases with time, and at equilibrium, it reaches a maximum. From then on, the entropy does not change with time.

6.2 Critique of the H-theorem, and Boltzmann's answers

This theorem drew great amount of criticism, the most well-known are:

I. The "Reversal Paradox":

The H-theorem singles out a preferred direction of time. This is inconsistent with the time reversal invariance of the equations of motion.

In other words, one cannot obtain time-asymmetric behavior, from time-symmetric equations of motion.

II. The "Recurrence Paradox", Based on Poincare's Theorem:

After a sufficiently long time, an isolated system with fixed E, V, N, will return to an arbitrary small neighborhood of almost any given initial state.

If we assume that $dH/dT < 0$ at all t, then obviously H cannot be a periodic function of time.

Both paradoxes have been with us ever since. Furthermore, most popular science books identify the Second Law, or the behavior of the entropy with the so-called *arrow* of *time*. Both paradoxes seem to arise from the conflict between the *reversibility* of the equations of motion on one hand, and the apparent *irreversibility* of the Second Law, namely that the H-function decreases monotonically with time, on the other hand. Boltzmann rejected the criticism by claiming that H does not always decrease with time, but only with high probability. The irreversibility of the Second Law is not absolute, but only highly improbable. The answer to the recurrence paradox follows from the same argument. Indeed, the system can return to the initial state. However, the recurrence time is so large that this is never observed, not in our lifetime, not even in the life time of the universe.

6.3   Where did Boltzmann and his critics go wrong?

Notwithstanding Boltzmann's correct answers to his critics, Boltzmann and his critics made an enduring mistake in the interpretation of the H-function, a lingering mistake that has hounded us ever since. This is the very identification of the function $-H(t)$ with the behavior of the entropy. This error has been propagated in the literatures to this day.

At this stage it is important to make the following distinction between the reversal of the system to its initial state, and the reversal of the entropy. As we have seen in the example of Figures 4.18 and 5.4, the system can always visit any state, including the initial state (state b in the figures but not a). In this sense we cannot claim that the process is irreversible. Boltzmann was right in claiming that reversal of the states is not impossible but highly improbable. Boltzmann was also right in claiming that the $H(t)$ function can go up and can fluctuate even after reaching an equilibrium state. In fact, the H function can even reach the initial value.

Unfortunately, Boltzmann was wrong in believing that the entropy can go down; "not impossible, but highly improbable." The entropy of the system like the *summit* of the hill in Figure 4.2, it cannot go down, nor reach the initial state. In order for the entropy to decrease to its value at the initial state, the system must visit the initial state, *and stay there* at equilibrium. See Figure 4.18a and 5.4a).

It is clear from the very definition of the function $H(t)$, that $-H(t)$ is a SMI, and if one identifies the SMI with entropy, then we go back to Boltzmann's identification of the function $-H(t)$ with entropy.

Fortunately, thanks to the recent derivation of the *entropy function*, i.e. the function, $S(E, V, N)$, based on the SMI, it becomes crystal clear that the SMI is not entropy!

Translating our findings in chapter 3 to the H-theorem we can conclude that $-H(t)$ is SMI based on the velocity distribution. Clearly, one cannot identify $-H(t)$ with entropy. To obtain the entropy one must first define the $-H(t)$ function based on the distribution of both the locations and momentum, i.e.

$$-H(t) = -\int f(\mathbf{R}, \mathbf{p}, t) \log f(\mathbf{R}, \mathbf{p}, t) d\mathbf{R} d\mathbf{p} \qquad (6.4)$$

This is a proper SMI. This may be defined for a system at equilibrium, or very far from equilibrium. To obtain the entropy we must take the maximum of $-H(t)$ over all possible distributions $f(\mathbf{R}, \mathbf{p}, t)$.

$$Entropy = \max_{over\ all\ fs} [-H(t)] \qquad (6.5)$$

We also believe that once the system attains an equilibrium, the $-H(t)$ attains its maximum value, i.e. we identify the *maximum* over all possible distributions with the maximum of SMI in the limit $t \to \infty$, i.e.

$$Entropy = \lim_{t \to \infty} [-H(t)] = Max\ SMI\ (at\ equilibrium) \qquad (6.6)$$

At this limit we obtain the entropy (up to a multiplicative constant), which is clearly not a function of time!

Boltzmann, as well as his critics and many others believed that $-H(t)$ is entropy, and the time dependence of $-H(t)$ is the same as the (supposed) behavior of entropy, namely; that entropy always increases with time. This belief rests on the *form* of the H-function which resembles the form of the Gibbs entropy, as well as what is commonly referred to as Shannon's entropy.

Thus, once it is understood that the function $-H(t)$ is an SMI and not entropy, it becomes clear that the criticism of Boltzmann's H-Theorem is addressed to the evolution of the SMI, and not to the entropy. At the same time, Boltzmann was right in defending his H-theorem when viewed as a theorem on the evolution of SMI, but he was wrong in his interpretation of the quantity $-H(t)$ as entropy.

The Boltzmann H-theorem and the enormous misconception about entropy and the Second Law that followed is summarized in Figure 6.1.

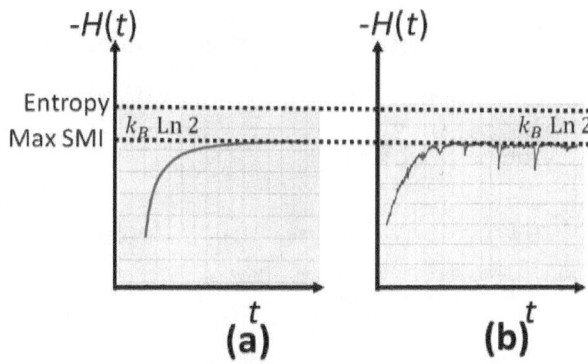

Figure 6.1. Schematic description of
(a) the implication of the H Theorem (equations 6.2 and 6.3)
and (b) the real variation of the H as a function of time.
The entropy of the system is constant, unchanging with time.

From equations 6.2 and 6.3, one can conclude that the function $-H(t)$ is a monotonic increasing function of time, and at $t \rightarrow \infty$ it reaches a constant value. See Figure 6.1.

Boltzmann, as well as many others believed that the H-theorem represents the behavior of entropy, namely that S increases monotonically with time until reaching a maximum value.

Critics of Boltzmann's theorem also believed that the entropy changes monotonically with time, and found a conflict between the H-theorem, on one hand, and the "time-reversal" of equations of motion, on the other hand. Boltzmann correctly answered his critics by claiming that $-H(t)$ increases monotonically only with high probability, and that at equilibrium the entropy might decrease; it is not impossible, but highly improbable event. Figure 6.1b.

Boltzmann's answer was correct, for as long as it applies to $-H(t)$ as a SMI. Both Boltzmann, as well as his critics were wrong in interpreting $-H(t)$ as entropy. Entropy *does not* change with time. It does not decrease spontaneously, not even with small probability.

This conclusion followed from the derivation of entropy from the SMI, and it is the essence of the "Timeless Nature of Thermodynamics."

It should be noted that in popular science books one finds figures similar to 6.1b but applied to the "entire universe." This is of course, meaningless since the entropy of the universe is not definable.

The latter conclusion also dismisses Clausius' formulation of the Second Law. Thus, the entropy of a well-defined thermodynamic system is *timeless*. The entropy of the universe is not *timeless*; it is simply *meaningless*.

6.4 Conclusion

Boltzmann's contribution to understanding entropy and the Second Law is undeniable and unshakable. However, Boltzmann also contributed to the lingering misinterpretation of entropy and the Second Law. The first misinterpretation associates entropy with disorder.

Boltzmann was probably the first to associate entropy with disorder. Here are some quotations:

"... the initial state of the system...must be distinguished by a special property (ordered or improbable) ..."

"...this system takes in the course of time states...which one calls disordered."

"Since by far most of the states of the system are disordered, one calls the latter the probable states."

"... the system...when left to itself, it rapidly proceeds to the disordered, most probable state."

It is apparent from these quotations that Boltzmann never "equated" entropy with disorder as many others did. However, the "disordered" interpretation of entropy has lingered on to this day. Criticism of this interpretation has been published earlier, see Ben-Naim (2012).

Boltzmann's second failure is associated with his H-function and H-theorem. As we have noted in section 6.3, the H-theorem drew a lot of criticisms which were successfully repelled by Boltzmann himself. The main argument of defending his H-theorem is based on the idea that the Second Law is basically a law of probability. But Boltzmann, as many others failed to pinpoint the *subject* on which probability operated. In other words, the failure to distinguish between the two statements:

1. When we remove a constraint, the *state* of the system changes with high probability to a new equilibrium state.

2. When we remove a constraint, the *entropy* of the system changes with high probability towards a maximum.

Whilst the first statement is correct, and in fact applies to any thermodynamic system (isolated, isothermal, isothermal isobaric, etc.), the second is incorrect. First, the entropy formulation of the Second Law applies only to isolated systems, and second, the correct formulation is that the SMI, defined on the distribution of locations and momenta of all particles in an isolated system tends to a maximum at equilibrium, and at equilibrium the value of the MaxSMI is proportional to the entropy of the system.

It is ironic that Shannon named his quantity entropy, and by doing so he contributed to the immense confusion in both thermodynamics and information theory. However, from the two theorems proved by Shannon, it follows that the entropy (the thermodynamic entropy) of the system is attained at a distribution (of locations and velocities) which *maximizes* the SMI. We also know that the distribution of locations of ideal gases is uniform, and the distribution of velocities is the Maxwell-Boltzmann distribution. Therefore, we can identify the maximal value of the SMI with the value of the SMI at equilibrium. Since there is only a single distribution that maximizes the SMI, it follows that the entropy of a thermodynamic system has a unique value. It does not change with time, it does not reach a maximum value with time, and it does not fluctuate at equilibrium (neither with low or high probability).

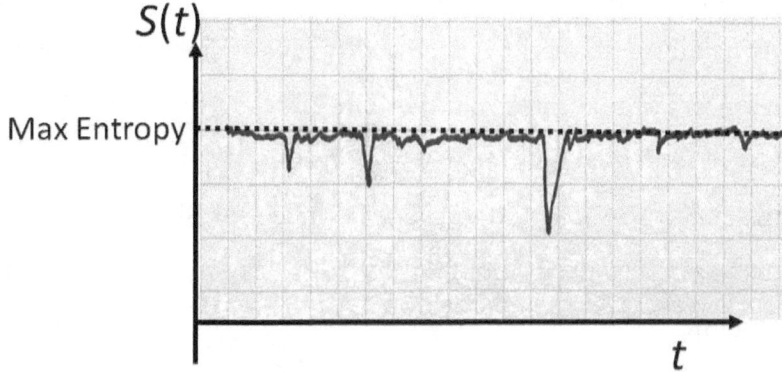

Figure 6.2. Schematic description of the entropy as a function of time for a system at equilibrium as described in the literature.

In many popular science books, one can find plots showing how entropy changes with time. See Figure 6.2. Most of the time fluctuations are small, but once in many billions of years it might have a big fluctuation. This behavior should be ascribed to the SMI, and not to the entropy.

Another misconception associated with time-dependence of entropy is associated with the so-called Past Hypothesis, namely that since entropy of the universe always increases, one can extrapolate back and conclude that the entropy of the universe at the Big Bang must have been very low. The second unwarranted "prediction" may be referred to as the Future Hypothesis which basically states that the universe is doomed to "thermal death." These two hypotheses are unwarranted. They were criticized in Ben-Naim (2012, 2017a, 2017c).

We can summarize the main conclusions of this book as follows:

1. It is essential to make a clear-cut distinction between the SMI and entropy. Failure to make such a distinction has caused great confusion in both thermodynamics and Information Theory.

2. To reserve and use the concept of entropy only to macroscopic systems at *equilibrium*, and to use the concept of SMI for all other systems; small or large number of particles, near or far from equilibrium.

3. Entropy, being a state function is independent of time.

4. The Second Law is basically a law of probability; systems will spend more time in states having larger probabilities. When the system contains a very large number of particles, "larger probability" becomes almost one, and "more time" turns into *always*.

5. Change of entropy (as well as in Helmholtz and Gibbs energies) is a *result* of the change in the state of the system, not the *cause* of that change.

6. As a result of removing some constraints in an isolated system, the entropy change will be either positive, or zero.

7. Entropy is proportional to the maximum SMI of a thermodynamic system which is the same as the SMI of the system at equilibrium.

8. Entropy may be defined and used without any reference to the Second Law. Likewise, the Second Law may be formulated and applied without ever mentioning entropy.

Notes

Note 1:   Clausius (1879) enunciated the First and the Second Law as:

"Die Energie der Welt ist Konstant.

Die Entropie der Welt strebt einen Maximum zu."

(the energy of the world is constant, the entropy of the world tends to a maximum)

Note 2:   Of course, probability *has no* natural tendency to increase. Events that have higher probability occur with high frequency. But probability, in itself, does not have any tendency to increase or decrease.

Note 3:   Clearly, the statement: "Entropy of the universe always increases" sounds more meaningful than "Entropy always increases." Unfortunately, both of the statements are meaningless. The universe is not a well-defined system. The entropy of the universe was never defined, and I doubt that it will ever be defined.

Note 4:   As I have discussed in Ben-Naim (2016a), it is far from clear that the metaphor of the Arrow of Time has any physical significance.

We say that time flows, but we really do not know whether time flows or not. I prefer to view "time" as an abstract one-dimensional and arrowless line along in which we order the occurrences of events. When we say that a certain amount of time has passed we actually *count* the number of times some periodic phenomenon has occurred. This can be the rotation of the earth about its axis, the periods of motion of a pendulum, or the period of vibration of a molecule. This *counting* is conventionally always a positive number. It is the same as the distance we travel when we go from A to B. The distance we travel is *doubled* when we go from A to B, then from B to A. The same is true when we count the total number of words, sentences, or books we write. This number always increases (or remains unchanged) even if we delete some words or burn some books. Leaving aside the question of the existence of the arrow of time, we can categorically say that entropy has nothing to do with time (with or without an arrow).

Note 5: Mackey's book starts with Eddington's (1935) statement:
"The law that entropy always increases, holds, I think, the supreme position among the laws of Nature."
Then the author says that:

"The central question this book addresses is the dynamic origin of the Second Law of Thermodynamics."

Thus, the book is not about the origins of "thermodynamic behavior," in general, but about the origin of the Second Law. From the preface we find:

"Let $S_{TD}(t)$ denote the thermodynamic entropy at time t."

As we shall see throughout the present book one cannot speak on the thermodynamic entropy (of what?) at time t, as much as one cannot talk about the beauty at time t.

In Chapter 1, Mackey identifies the H-function as the entropy (with a minus sign). This is one of the recent repetition of Boltzmann's error regarding the H-Function. See also Chapter 6 for more details.

Note 6:   When I say three definitions of entropy I do not mean three *different entropies*. What I mean by three definitions is that they lead to the same concept of entropy. More specifically, when we calculate changes of entropy based on Boltzmann's definition we get the same result as we get from the Clausius definition (this is true for all those changes of entropy that are calculable and therefore the results can be compared). Likewise, the calculated changes of entropy based on SMI agree with those calculated by either Boltzmann's or Clausius' definitions.

Note 7:   See Ben-Naim (2016b, 2017c).

Note 8:   See Ben-Naim (2016b, 2017c).

Note 9:   To the best of my knowledge no one has *defined* entropy using the SMI. Therefore, I will often refer to this definition as either the *definition based on SMS*, or simply as ABN definition of entropy. Originally, this definition was used as an interpretation of entropy based on the SMI [Ben-Naim (2006, 2007, 2008)].

Note 10: Note however that the "falling of water," and the "flow of heat" are governed by very different laws. The first is governed by Newton's gravitation law, the second, by the Second Law of Thermodynamics. It is unfortunate that many authors claim that the Second Law is the *cause* of everything that happens, including water falling.

Note 11: Denoting by $\Delta Q(Hot)$ and $\Delta Q(Cold)$ the amounts of heat flowing inside and outside the engine, respectively. The efficiency of the heat engine is defined by:

$$\eta = \frac{\Delta W}{\Delta Q(Hot)} = \frac{\Delta Q(Hot) - \Delta Q(Cold)}{\Delta Q(Hot)} = 1 - \frac{\Delta Q(Cold)}{\Delta Q(Hot)}$$

where ΔW denotes the amount of useful work done by the system (e.g., lifting a weight).

In the Carnot engine operating between the two temperatures T_2 and T_1 ($T_2 > T_1$) the efficiency is given by:

$$\eta = \frac{T_2 - T_1}{T_2} = 1 - \frac{T_1}{T_2} \leq 1$$

Since $T_1 < T_2$, $\eta < 1$. The efficiency is zero when $T_1 = T_2$, and it approaches one when $T_1/T_2 \to 0$.

Note 12: Sometimes you see a slightly different notation, i.e. dQ_{rev}/T, where "rev" is short for reversible. We will not need this notion here. We require that the system is large enough at a specific temperature T, and that dQ is small enough so that the heat transferred does not change the temperature of the system. Some authors use the notation; δQ, to emphasize that this quantity is not an exact differential. On the other hand, dS is an exact differential. This means that there exists a function S, which is a state-function, i.e. a function of the parameters describing the system, say T, P, N, and is differentiable with respect to these variables.

Note 13: This is discussed in details in Ben-Naim (2008).

Note 14:   This will be further discussed in Chapter 6. Here we only note that the system *can* get back to its initial state, but the entropy can *never* decrease spontaneously in isolated systems.

Note 15:   The unwary reader might be puzzled by this statement. Shannon himself referred to the SMI as entropy in 1948, so what makes this definition "recent?"

In this book we shall make a clear-cut distinction between the SMI (sometimes referred to as Shannon entropy, or informational entropy), and the thermodynamic entropy. The fact that Shannon chose to call his measure "entropy" was a grave mistake which caused great confusion in both information theory and in thermodynamics.

The first derivation of the *entropy function* for an ideal gas was published in Ben-Naim (2006). It was later extended to also include systems of interacting particles in 2008.

Note 16.   The reader might find this comment trivial and superfluous. It is clear that $Nk_B \ln 2$, or $R\ln 2$ (for one mole of gas) is independent of temperature. Yet, when I wrote this trivial fact in a paper I submitted to the Journal of Chemical Education, it was *rejected* because the reviewer claimed that entropy change in this process depends on temperature. The details of this story are mentioned in Ben-Naim (2012), where one can read the "origin" of the temperature dependence of the gas constant.

Appendix A: The many faces of reversibility and irreversibility

The Second Law is often referred to as the Law of irreversibility. On the other hand, the equations of motion of the individual particles are said to be "reversible." If that is so, then there is an apparent conflict between the behavior of a simple particle and the behavior of a system containing a very large number of particles. Unfortunately, the are several meanings to the term reversibility and irreversibility, and the whole conflict arises because of the confusion between the different types of reversibility.

Even in the context of thermodynamics there are at least four senses assigned to the term "reversible process." Here are the four senses that I can identify as distinct.

Rev1: The mechanical sense.

Rev2: A thermodynamic process for which the entropy change is zero.

Rev3: A thermodynamic process: $A \rightarrow B$ that proceeds along a dense series of equilibrium states.

Rev4: A thermodynamic process: $A \rightarrow B$ that can be reversed, i.e., B   A.

Before we discuss the four definitions of the term "reversibility" in thermodynamics, it is instructive to see that this term could have several different meanings when used colloquially.

Suppose a person walks from point A to point B. After that, you are told that the same person went back to point A. You might be wondering how this process was *reversed*.

Did the person walk backwards, as would have been seen by rewinding the movie showing the person going forward and backwards? This reversal of the motion would be fun to watch, but not a realistic process.

Another reversal process that one can imagine, but still almost unrealistic is that the person went back in such a way that after the completion of the cycle A→B→A everything in the universe had returned to the initial state. Clearly, this is not a realistic process, but still an imaginable one.

Another reversal, now more realistic is that the person simply went back from A to B along the same path. Although the same person retraced the same path, some changes in both the person and in the environment had necessarily occurred.

The fourth possibility is the simplest, the person returned to A, not necessarily along the same path, and of course changes must have occurred in both the person and in the entire universe.

Bearing these examples in mind let us move on to discuss the various meanings of "reversibility" as used in thermodynamics.

(i) The sense Rev1 is used in mechanics in connection with the reversibility of the equations of motion. A process is said to be reversible if by inversion of the velocities of all particles, the process is reversed. This sense of the term is not usually used in thermodynamics. It is sometimes used in statistical mechanics in connection with the apparent conflict between the reversibility of the equations of motion of the atoms and molecules, and the irreversibility (in the sense discussed in (iv) below) of the spontaneous thermodynamic process.

(ii) The sense Rev2 is sometimes used in connection with the Second Law of thermodynamics. It states that in any spontaneous process occurring in an isolated system, the entropy can never decrease. A process for which the entropy increases is called irreversible process. A process along a constant-entropy path is referred to as a reversible process. This nomenclature is used by Callen (1985) to distinguish between reversible and quasi-static processes. For instance, removing a partition separating two different gases will result in a spontaneous mixing and the entropy increases. This specific process is deemed to be irreversible.

(iii) The sense Rev3 is most commonly used in thermodynamics. It is sometimes confused with the sense Rev2. The difference between the two can be clarified by the following simple example.

Consider a spontaneous process of expansion. Initially, all the N molecules are confined to one compartment. Removal of the partition causes a spontaneous process of expansion. This is a typical irreversible process in the sense discussed in (ii). The initial and the final states of the system may be described by two points in the *PV* diagram.

Clearly, the process is not reversible in the sense Rev2. But it is also not reversible in the sense Rev3 since we cannot trace back the path of the process in the *PV* diagram. This is so simply because the thermodynamic states of the system *during* the expansion process are not well defined, and therefore it is meaningless to talk about reversal along exactly the same thermodynamic path from the initial to the final state.

Next, suppose that we change the weight on the piston gradually, each time reducing the weight by dM, waiting for the system to reach an equilibrium state, then reducing the weight again by dM, and so on. In this process we can draw all the equilibrium points in the *PV* diagram. We can say that we know, say, ten points on the path from the initial to the final states, but we do not know the exact path between any two equilibrium points.

Next, we can imagine that at each step we remove only an infinitesimal weight dM on the piston. In the limit, when we perform this process in infinitesimal steps, we can draw an almost continuous curve in the *PV* diagram. Clearly, in this limiting process there is a *path* in the *PV* diagram leading from the initial to the final states. Therefore, it is meaningful to speak about the thermodynamic path from the initial to the final state, as well as about the reversed path from the final to the initial state.

To distinguish between this reversible process from the reversible process in the sense Rev2, the former is sometimes referred to as a quasi-static process. A quasi-static process is simply a process which is carried out in very small steps so that effectively the system goes through an almost continuous series of equilibrium states. Because each equilibrium state is well-defined thermodynamically, it follows that the path of the process is well-defined. It is therefore meaningful to talk about reversing the path or reversing the process.

It should be stressed that a quasi-static process, i.e. a reversible process in the sense Rev3 does not imply that the process is reversible in the sense Rev2. It is therefore advisable to use two different terms to distinguish between the two.

Note also that reversible in the sense Rev3 is not reversible in the sense Rev1. The thermodynamic path is reversed, but the molecular trajectories are not.

(iv) The weakest form of reversibility is that a process A→B can be reversed B→A. There is no requirement that the reversal should be along the same thermodynamic path (sense Rev3), or that there is no change in entropy (sense Rev2). It is the weakest since it is difficult to find a thermodynamic process that cannot be reversed in this sense. We exclude from this discussion processes of life and death which, at least at the present level of our knowledge seems to be completely irreversible, in its colloquial sense. An example which is given quite often is boiling an egg. Such a process cannot be reversed (un-boil the egg?) However, in thermodynamics we are discussing processes from one well-defined *equilibrium state* to another *equilibrium* state. It is far from clear whether an egg is in an equilibrium state either before or after the cooking. The boiling of an egg like any other spontaneous processes are "irreversible" only in a statistical sense.

Appendix B. On the validity of the main assumption of Non-equilibrium thermodynamics

Introduction

Non-equilibrium thermodynamics is founded on the assumption of *local equilibrium*.

Typically, this assumption states that (de Groot, S. R., and Mazur, P. (1962):

"It will now be assumed that, although the total system is not in equilibrium, there exists within small elements a state of 'local' equilibrium for which the local entropy s is the same function of u, v and c_k as in real equilibrium."

No justification is provided for this assumption, nor a proof that such a "local entropy" would have the "same function of u, v and c_k as in real equilibrium."

Most textbooks on non-equilibrium thermodynamics starts with the reasonable assumption that in such systems the intensive variables such as temperature T, pressure P, and chemical potential μ may be defined operationally in each small element of volume dV of the system.

Thus, one writes:

$$T(\mathbf{R}, t), \ P(\mathbf{R}, t), \ \mu(\mathbf{R}, t) \tag{B.1}$$

where $\mathbf{R}$ is the locational vector of a point in the system, and t, is the time. One can further assume that the density $\rho(\mathbf{R}, t)$ is defined locally at point

R and integrate over the entire volume to obtain the total number of particles N

$$N = \int \rho(R, t)dR \tag{B.2}$$

Similarly, one can define densities $\rho_k(R, t)$ for each component k, of the system.

One can also define the internal energy per unit of volume $u(R, t)$. It is not clear however, how to integrate $u(R, t)$ over the entire volume of the system to obtain the total internal energy of the system. While this may be done exactly for ideal gases, i.e. when the total energy of the system is the sum of all the kinetic (as well as internal) energies of all the particles in the system, it is not clear how to do the same for systems having interacting particles for which one defines the integral:

$$U = \int u(R, t)dR \tag{B.3}$$

Here, the integration is essentially the sum over all the $u(R, t)$, each of which is defined in a small cell $dR(dxdydz)$, or dV in the system, neglecting the interaction energies between the different cells. When there are interactions between the particles, it is not clear how to account for these interactions in the very definition of the local energy, u (**R**, t). Similarly, it is far from clear whether one can write the entropy of the entire system as an integral of the form:

$$S = \int s(R, t)dR \tag{B.4}$$

The most important and unjustified assumption is related to the definition of the local entropy $s(R, t)$. One assumes that the local entropy function, $s(u, v, n)$ is the same as the function $S(U, V, N)$, i.e. s is the same function of the local energy, volume, and number of particles of each element of volume.

This assumption may be justified for ideal gas when the distribution of locations and velocities is meaningful for each element of volume in the system. To the best of the author's knowledge this assumption has never been justified for systems of interacting particles. The main difficulty is that for such systems there are correlations between the elements of volumes, which, in turn, adds mutual information to the Shannon measure of information (SMI) of the system, hence also on the entropy of the system. When these correlations are large, it is not clear whether one can define the local entropy, and therefore the validity of the integral is questionable.

Once one makes such an assumption, one writes the changes in the entropy of the entire system as:

$$dS = d_e S + d_i S \qquad (B.5)$$

where $d_e S$ is the entropy change due to the heat exchange between the system and its surrounding, and $d_i S$ is the entropy produced in the system. For an isolated system $d_e S = 0$, and all the entropy change is due to $d_i S$. The latter is further written as:

$$\frac{d_i S}{dt} = \int \sigma d\boldsymbol{R} \qquad (B.6)$$

where $\sigma(\boldsymbol{R}, t)$ is referred to the *local* entropy production:

$$\sigma(\boldsymbol{R},t) = \frac{d_i s}{dt} \geq 0 \qquad (\text{B.7})$$

Thus, for an isolated system one has a local entropy production which is a function of time, and after integration one also obtains the total change of the entropy of the system as a function of time. Since the quantity σ is defined in terms of the local entropy function $s(\boldsymbol{R},t)$, and since $s(\boldsymbol{R},t)$ is not a well-defined quantity, one should doubt the whole theory based on the assumption of local equilibrium.

In the following sections, we will show that:

1. It is *not clear* how to *define* a local entropy density function $s(\boldsymbol{R},t)$.
2. It is *almost certain* that if one could define an entropy density, it would not have the same functional dependence on the local quantities (u,v,n) as the entropy of the whole system depends on the macroscopic variables (U,V,N).
3. It is *not true* that such an entropy-density function $s(R,t)$ could be integrated to obtain the entropy of the entire system.

How to define a local entropy-density function?

When a system is at equilibrium, the entropy is a homogenous function of its extensive variables, i.e. for any $\lambda > 0$ one can write:

$$S(\lambda U, \lambda V, \lambda N_1, \cdots, \lambda N_c) = \lambda S(U,V,N_1,\cdots,N_c) \qquad (\text{B.8})$$

One can use this property to *define* a local entropy-density simply by choosing $\lambda = V^{-1}$, i.e.

$$S = S/V = s\left(\frac{U}{V}, \frac{N_1}{V}, \cdots, \frac{N_c}{V}\right) \tag{B.9}$$

where N_i/V is the number density of the ith component. Clearly, at equilibrium, s has the same value in any element of volume dV at a point $\boldsymbol{R}$ in the same system. Therefore, one can integrate over the volume of the system to obtain the total entropy of the system:

$$\int s(\boldsymbol{R})dV = s(\boldsymbol{R})V = S \tag{B.10}$$

This procedure would not work for a system which is not at equilibrium. The reason is that in a non-equilibrium system we do not have the entropy of the entire system to begin with. Therefore, we cannot define the entropy-density as in Eq. (B.9).

For a non-equilibrium system, we have to define first the local entropy-density then integrate to obtain the total entropy of the system.

Let us examine whether this can be done by using the following three definitions of entropy.

(i)   *Using Clausius' definition*

Clausius originally defined a small change of entropy dS due to a small addition of heat dQ to a system at constant temperature. Implicit in this definition is the assumption that while we transfer dQ, the system's temperature does not change. This assumption is valid for a *small* addition of heat dQ, to a *macroscopic* system at a constant temperature. It would not work for a very small system. Even if one can define the local temperature $T(\boldsymbol{R})$, the addition of dQ to an infinitesimal volume dV would cause a very big change in the local temperature.

(ii)   *Using Boltzmann's definition*

It is well-known that starting with the Boltzmann definition of entropy $S = k_B lnW$ (k_B being the Boltzmann constant, and W the number of accessible microscopic states of the system), one can derive an explicitly entropy function $S(U, V, N)$ for an ideal gas (of one component):

$$S(U,V,N) = Nk_B \ln\left[\frac{V}{N}\left(\frac{4\pi mU}{3h^2N}\right)^{3/2}\right] + \frac{5}{2}Nk_B \qquad (B.11)$$

This is the Sackur-Tetrode equation (where m is the mass of the particles, h is the Planck constant).

Clearly, the function $S(U, V, N)$ has the property of being a homogenous function of the first order. However, in the derivation of the Sackur-Tetrode equation one assumes that the number of particles N is large enough so that the Stirling approximation may be applied to $N!$, i.e.

$$\ln N! \approx N \ln N - N \qquad (B.12)$$

For small elements of volume, the number of particles in each element might be so small that the approximation no longer applies.

This is certainly true for infinitesimal volumes in which the number of particles will either be zero or one. If one follows the derivation of Sackur-Tetrode equation and stops before the introduction of the Stirling approximation, one gets the equation:

$$S \sim k_B \left[N \ln \left[V \left(\frac{2\pi e m k_B T}{n^2} \right)^{3/2} \right] - k_B \ln N! \right. \tag{B.13}$$

Clearly, for N of the order of 1 this "entropy" will not be a homogenous function of order one is N. Therefore, an entropy-density function will not have the same dependence on (u, v, n) as the entropy function $S(U, V, N)$.

(iii)   Using the SMI-based definition

The definition of entropy based on the Shannon measure of Information (SMI)[12] leads to the Sackur-Tetrode equation for an ideal gas of simple particles[7-11]. Therefore, the conclusion reached based on the Boltzmann definition will be the same as that reached by the definition based on the SMI.

Thus, we can conclude that whatever definition is used for the entropy of a macroscopic system, it will not apply to an infinitesimally small system. Therefore, one cannot assume that the local entropy density will have the same dependence on (u, v, n) as the function $S(U, V, N)$.

Can one integrate the local entropy density to obtain the entropy of the system?

Suppose that a local entropy-density $s(\boldsymbol{R}, t)$ can be defined, such that $s(\boldsymbol{R}, t)dV$ is the entropy of a small element of volume dV, at time T. The question is whether one may integrate such a function to obtain the entropy of the entire system, $S(t)$, i.e.

$$S(t) = \int_V s(\boldsymbol{R}, t)dV \tag{B.14}$$

We examine the question of integrability by starting with two independent systems A and B, each being a macroscopic system at equilibrium. In such a case, the additivity property of the entropy applies, i.e.

$$S(A + B) = S(A) + S(B) \qquad (B.15)$$

Such an additivity applies to any number of independent systems, i.e.

$$S(total) = \sum_{i=1}^{c} S(i) \qquad (B.16)$$

where $S(i)$ is the equilibrium thermodynamic entropy of system I, Figure A.1.

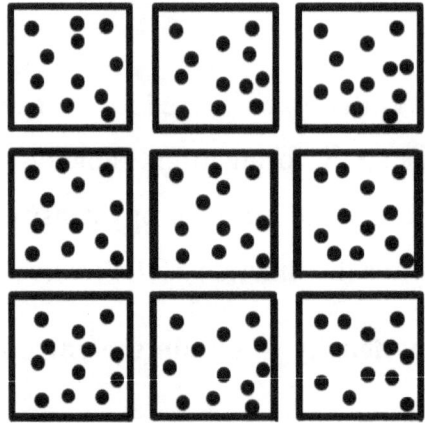

Figure A.1. Six isolated systems, each having the same *energy, volume and number of particles* at equilibrium.

Next, we bring the systems into interacting-contact. We imagine that each system is an isolated system with a well-defined entropy $S(i)$. We assume that the walls are impermeable to heat, to volume, and to particles but they are so thin that particles in i can interact with particles in j, where i and j are adjacent systems, Figure A.2.

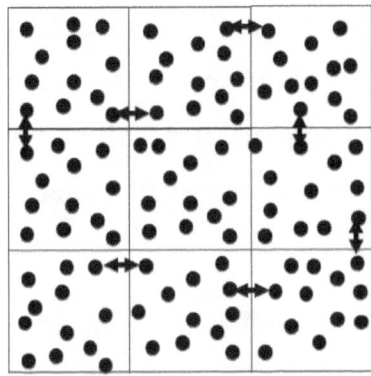

Figure A.2. Six isolated systems, each having the same *energy, volume and number of particles*. The particles can interact across the walls. Some of the interactions are shown by a double arrow

Such idealized walls do not exist, and we shall soon remove these. However, assuming the existence of such walls, we could write the total entropy of all the c-systems as:

$$S(Total) = \sum_{i=1}^{c} S(i) + \sum_{i=1}^{c} \sum_{j=1}^{c} S_{in}(i,j) \qquad (B.17)$$

The double summation is for al and j, such that $i<j$. As before $S(i)$ is the (macroscopic) entropy of the ith system, and $S_{in}(i,j)$ is the entropy due to the intramolecular interaction between particles in j, and particles in i. These entropies of interactions can be cast in the form of mutual information between any pairs of systems.

When all the systems are macroscopic one can neglect all the interaction entropies between the systems. This assumption follows from the fact that only particles on the surface of i can interact with particles on the surface of j. Since the number of particles on the surface of each system is negligible (for macroscopic systems) compared with the number of interacting particles within each system we can neglect the mutual information between any pair of system (hence, the entropy of interaction) compared to the mutual information due to intermolecular interaction within each system. Such an approximation will preserve the additivity property (B.16) in spite of the existence of interactions between the systems.

Clearly, this approximation becomes less and less valid when the systems become smaller. In the limit when the "systems" are infinitesimally small, the interactions between particles in different elements of volume ("system") will dominate the entropy due to interactions in the entire system.

When we can neglect high order correlation functions for non-ideal gases, the mutual information (MI) in the entire system (hence, the entropy due to intermolecular interactions) has the form:

$$MI = -\frac{N(N-1)}{2} \int P(R_1, R_2) \log[g(R_1, R_2)] dR_1 dR_2 \qquad (B.18)$$

where $P(R_1, R_2) dR_1 dR_2$ is the probability of finding a specific particle in dR_1 at R_1, and another specific particle in dR_2 at R_2.

It is not clear how to divide this total mutual information into a sum (or an integral) over all the elements of volume dV. Specifically, when each element of volume can accommodate almost one particle, hence, such an element of volume would not have any mutual information (hence, no entropy due to interactions).

The situation is far more serious in a condensed phase when higher order correlations become significant. As is well-known even when intermolecular interactions are strictly pairwise, higher order correlations are not negligible. This in turn means that instead of (B.18), we must include mutual information between all particles in the system, i.e.:

$$\text{MI} = -\int P(\boldsymbol{R}^N)\log[g(\boldsymbol{R}^N)]dR^N \tag{B.19}$$

where

$$g(\boldsymbol{R}^N) = \frac{P(\boldsymbol{R}^N)}{\prod_{i=1}^N P(\boldsymbol{R}_i)} \tag{B.20}$$

This mutual information is a property of the entire system and cannot be "divided" into contributions due to small, or infinitesimal elements of volume in the system.

Conclusion

To the best of our knowledge the assumption of "local equilibrium," followed by the definition of entropy-density function in a system far from equilibrium has never been founded either within thermodynamics, or from statistical mechanical arguments.

In this Appendix we argued that the very assumption of "local-equilibrium" cannot be justified for systems far from equilibrium.

Furthermore, we argued that an entropy density function, even if it could be defined would not have the same functional dependence on the local energy, volume, and number of particles (u, v, n) as the entropy of a macroscopic system at equilibrium $S(U, V, N)$. We also showed that for condensed systems where intramolecular interactions are significant, one cannot derive an entropy function for the whole system by integrating over the *local* entropy-density function.

These findings shed serious doubts on the thermodynamics of systems far from equilibrium.

In the rest of this Appendix a few quotations from the literature will be presented:

1.   Glansdorff and Prigogine (1971) on page 12 start with the division of dS into two terms:

$$dS = d_iS + d_eS$$

where d_eS is the change in entropy due to the flow of heat (i.e. $dS = \frac{dQ}{T}$), and d_iS is the "entropy production due to changes inside the system."

Then, they write the equation:

$$\frac{d_iS}{dt} = \int \sigma[S]dV \geq 0$$

Calling $\sigma[S]$ the "entropy source," the entropy production per unit time, and volume"

Then on page 14:

"This will be the case when there exists within each small mass element of the medium a state of local equilibrium for which the local entropy s is the same function of the local macroscopic variables as at equilibrium state. This assumption of local equilibrium is not in contradiction with the fact that the system as a whole is out of equilibrium."

No justification is provided for the assumption of local equilibrium.

2. de Groot and Mazur (1962) start from the formulation of the Second Law as:

$$dS \geq 0 \text{ (for an adiabatically insulated system)}$$

Then introduce the entropy s per unit mass defined by the equation:

$$S = \int_V \rho s dV$$

where ρ is the mass density (mass per unit volume) ρ_k for the component k. They write:

"It will now be assumed that, although the total system is not in equilibrium, there exists within small elements a state of "local" equilibrium, for which the local entropy s is the same function (14) of u, v, and c_k as in real equilibrium."

Thus, the assumption is made that the "local entropy" s is the same function of (u, v, c_k) as in real equilibrium system.

No justification for this hypothesis of "*local equilibrium*" is provided. They claim that this can be justified by virtue of the validity of the conclusions derived from it.

3. Callen (1985), in Chapter 14 on "Irreversible Thermodynamics"

"One problem that immediately arises is that of defining entropy in a nonequilibrium system. This problem is solved in a formal manner as follows:

To any infinitesimal region we associate a local entropy $S(X_0, X_1, ...)$, where, by definition, the functional dependence of S on the local extensive parameters $X_0, X_1, ...$ is taken to be identical to the dependence in equilibrium.

Again, the local intensive parameter F_k is taken to be the same function of the local extensive parameters as it would be in equilibrium. It is because of this convention, incidentally, that we can speak of the temperature varying continuously in a bar, despite the fact that thermostatics implies the existence of temperature only in equilibrium systems.

The rate of local production of entropy is equal to the entropy leaving the region, plus the rate of increase of entropy within the region."

I doubt that the introduction of the "local equilibrium" assumption "solves" the problem.

4. Kondepudi and Prigogine (1999), in Modern Thermodynamics, on pages 6-7 introduce the energy density $u[T(x), n(x)]$ defined as the *"internal energy per unit volume."* Then the total internal energy U is obtained as:

$$U = \int_V u[T(x), n_k(x)]dV$$

where the integration is carried over the entire volume of the system.

$n_k(x)$ is the number of moles of the kth component per unit volume at point x.

Then, they write:

$$N = \int_V n_R(x)dV$$

This should be N_k, the total number of molecules of type k. While N_k can be defined as a sum or integral over all the elements of volumes in the system, the equation for the energy is valid only when the local $T(x)$ is well-defined at each point x, and that neglects interaction between the different elements of volume. This assumption is not discussed and not justified. Similarly, they define the entropy density $s[T(x), n(x)]$, and relate it to the total entropy of the system:

$$S = \int_V s\,[T(x), n(x)]dV$$

If there are interactions between the particles, it is not clear how one *defines* the entropy density, and it is far from clear how one *defines* the entropy density, and it is far from clear that the entropy of the system can be written as an integral over the entropy densities.

It is important to note that the authors *do not* define $s[T(x), n(x)]$! They only say that "Similarly, an entropy density, $s(T, n_k)$ can be defined" without actually defining it. Clearly, the integral above does not *define* either the entropy of the system, or the entropy density $s(T, n_k)$.

On page 7, the authors comment:

"In texts on classical thermodynamics, when it is sometimes stated that the entropy of a non-equilibrium system is not defined, it is meant that S is not a function of the variables U, V, and N. If the temperature of the system is locally well defined, then indeed the entropy of a non-equilibrium system can be defined in terms of an entropy density, as in (1.2.3)."

This comment is doubly misleading. First, when people say that entropy of a non- equilibrium system is not defined, they mean *it is not defined*. Second, even if the temperature of the system is locally well defined (I doubt that this can be defined at any arbitrary small element of volume), it does not follow that the equation for the entropy above, defines the entropy of the system in terms of an undefined entropy density.

5. In a more recent textbook by Kondepudi (2008), we find the explicit assumption made about the *local equilibrium*.

"The basis of the modern approach is the notion of local equilibrium. For a very large class of systems that are not in thermodynamic equilibrium, thermodynamic quantities such as temperature, concentration, pressure, internal energy remain well-defined concepts locally, i.e. one could meaningfully formulate a thermodynamic description of a system in which intensive variables such as temperature and pressure are well defined in each elemental volume, and extensive variables such as entropy and internal energy are replaced by their corresponding densities. Thermodynamic variables can thus be functions of position and time. This is the assumption of local equilibrium."

Although I agree that local intensive quantities such as temperature, pressure, and densities may be defined operationally, I doubt that this can be done for either the internal energy or the entropy. No justification for this assumption is provided. Although the author admits that his assumption might not be a good approximation for some systems, he does not explain for which systems the approximation is valid.

6. Kreuzer (1981), on page 3 of his book writes:

"To explicitly state the conditions under which the assumption of local equilibrium is valid, the methods of nonequilibrium statistical mechanics are required. So far, this has been done rigorously and explicitly only for a dilute gas."

Indeed, if the system is an ideal gas, or a real gas but with negligible interaction energy, then one can express the total energy of the system as a sum or an integral over all infinitesimal values.

Appendix C: A few quotations on Time and the Second Law

In this Appendix we will present some quotations from the relatively recent literature on the relationship between time and Entropy and the Second Law.

We will start with Albert (2000), who coined the term "Past Hypothesis" which I believe is responsible for much of the confusion about the "entropy of the Universe."

In the preface the author makes the following statements regarding the aim of the book:

"Elementary introduction and as an original contribution to the development of a scientific account of the distinction between the past and the future."

Later, the author makes it more explicit that the *"Second Law of thermodynamics, which is the point at which distinctions between past and future have made their most explicit...and most intensely studied appearance in the law of physics."*

In my opinion, the entire book of Albert: "Time and chance" is misleading. There is no law of physics, including the Second Law which distinguishes between the past and the future.

Furthermore, the author confuses the "reversibility" of the Newtonian equations of motion with the apparent "irreversibility in physical processes that do not happen backwards."

In Chapter 4, the author repeats the (erroneous) idea that the "laws of thermodynamics have a temporal direction in them." Then, the author draws several possible functions of entropy as a function of time (similar to the Figure 6.2. These are all pure meaningless figures.

From the "fact" that entropy always increases the author concluded that the universe must have started with a *low entropy state.*

From this erroneous (actually meaningless) assumption he arrives at the so-called "Past Hypothesis."

Davies (1977) discussed in great detail the question: "What is the cause of change in the universe?" then claims that "Boltzmann had proved that entropy could only increase." Then, he shows a figure (3.4 on page 73) where entropy can go down, but this is a very rare event, adding, "On exceeding rare occasions, a large fluctuation occurs and the entropy drops dramatically."

See my discussion in chapter 6 and Figure 6.2.

In Sachs (1987) book, on "The Physics of Time Reversal," the *time reversal transformation* T is expressed as:

$$T: t \rightarrow t' = -t$$

Read: "Under T, t transforms to $t' = -t$.

Then, he states: "It will be recognized that this is an improper transformation, a reflection akin to P. For the velocity it follows that:

$$T: v \rightarrow v' = -v$$

This transformation may be said to reverse the velocity. T is often referred to as "motion reversal" rather than "time reversal."

"…equations of motion are unchanged under the transformation T. Thus, the acceptance of a convention led naturally to an implicit assumption of time reversal invariance $(T = invariance)$ despite the apparent irreversibility of nature."

On page 26-27, Sachs discusses the H-Theorem, saying that Boltzmann's critics argued that the H-theorem led to a "paradox." Boltzmann used time-invariant equations of motion to derive a result that violated the time invariance. Then, concludes that the H-theorem "implies the inevitable increase of entropy towards equilibrium, is a "proof" of the Second Law of Thermodynamics for the perfect gas."

This is not true. Boltzmann *did not* prove the Second Law, and what he did does not apply to "ideal gases."

Wehrl (1991), in an article entitled "The Many Facets of Entropy," writes:

"Whereas the thermodynamic entropy refers to equilibrium states, the Boltzmann entropy [meaning $-H(t)$] is also defined in the non-equilibrium situation."

While I agree that entropy is defined for equilibrium states, I do not agree with the view that Boltzmann's entropy (meaning $-H(t)$) gives "precise meaning to the phenomenon of irreversibility." Instead, I suggest to keep the term entropy for equilibrium system and refer to the -H function as a SMI.

Lebowitz (1993), in his article on "Boltzmann's Entropy and Time's Arrow," summarized the problem as follows:

"Given that microscopic physical laws are reversible; why do all macroscopic events have a preferred time direction? Boltzmann's thoughts on this question have withstood the test of time."

Then, he explains:

"However, unlike S_G, which does not change in time even for ensembles describing the (isolated) systems not in equilibrium, S_B typically increases in a way that explains the evolution toward equilibrium of such systems."

In my view, there is only *one* thermodynamic entropy, and this entropy, does not change with time. What is referred here as Boltzmann's entropy (S_B) is nothing but SMI, multiplied by a constant.

Lebowitz (1999), in his article on "Microscopic Origins of Irreversible Macroscopic Behavior," starts his article with the statement:

"Time-asymmetric behavior as embodied in the second law of thermodynamics is observed in individual macroscopic systems."

Then states: *"I will try to clarify a time symmetric description of the dynamics of atoms to a time asymmetric description of the evolution of macroscopic systems."*

In my view, the "time-asymmetric" behavior of the macroscopic system is only apparent, and the Second Law is not "Time-asymmetric."

Uffink, J. (2001) wrote an exceptionally clear and courageous article entitled:

"Bluff your way in the Second Law of Thermodynamics."

In the abstract to the articles, the author concludes:

"I therefore argue for the view that the Second Law has nothing to do with the arrow of time."

I wholeheartedly agree with Uffink's conclusion. I would also add that I am not sure whether an "Arrow of Time" has any physical meaning. Therefore, my conclusion is that either entropy or the Second Law have nothing to do with *time* (with or without an arrow).

Carroll (2010) dedicates almost an entire book to the Past-hypothesis, the low entropy of the universe at, or near the Big-Bang and many other nonsense ideas. (For details, see Ben-Naim, 2016 and 2018b). Commenting on the Past Hypothesis he says:

"We must also assume that the observable universe began in a state of very low entropy. David Albert has helpfully given this assumption a simple name: the Past Hypothesis."

Not only this, Carroll says on page 43:

"When it comes to the past, however, we have at our disposal both our knowledge of the current macroscopic state of the universe, plus the fact that the early universe began in a low-entropy state. That one extra bit of information, known simply as the "Past Hypothesis" gives us enormous leverage when it comes to reconstructing the past from the present."

The entropy of the universe is not defined neither in the present, nor in the future. The statement, that "the fact that the early universe began in a low entropy state," is totally meaningless. The claim that this (meaningless) bit of information gives us an "enormous leverage" when it comes to reconstructing the past from the present is *silly, meaningless and an absurdity*!

Müller (2016) in his recent book writes:

"Eddington attributed the flow of time to the increase in entropy, a measure of disorder in the universe. We now know enormously more about the entropy of the universe than did Eddington in 1928 when he proposed the theory, and I'll argue that Eddington got it backward. The flow of time causes entropy to increase, not the other way around."

It is true that Eddington associated the so-called Time's arrow with entropy.

However, I do not agree with several statements made by Müller:

First, entropy is not "a measure of disorder in the universe." It is strange to read such statement in a book published in (2016).

Second, we do not know anything about the entropy of the universe. Therefore, the statement about our knowledge of the entropy of the universe now, or in 1928, is highly misleading. We know *nothing*, absolutely nothing on the entropy of the universe.

Third, I do not agree that "the flow of time causes entropy to increase, not the other way around." This is a doubly misleading statement. The entropy, in itself, cannot be said to increase or decrease. Entropy is a state function, and as such it is *defined* for a well-defined system, and as such it is not a function of time; the flow of time is *not* the cause of entropy increase, nor the other way around!

References and suggested reading

Albert, D.Z. (2000), *Time And Chance*, Harvard University Press, London

Atkins, P. (2007), *Four Laws that Drive the Universe*, Oxford University Press.

Atkins, P. (1984), *The Second Law*. Scientific American Books, W. H. Freeman and Co., New York.

Ben-Naim, A. (1987), *Is Mixing a Thermodynamic Process?* Am. J. Phys. **55**, 725.

Ben-Naim, A. (1992), *Statistical Thermodynamics for Chemists and Biochemists*. Plenum Press, New York.

Ben-Naim, A. (2006), American Journal of Physics. **74**, 1126.

Ben-Naim, A. (2007), *Entropy Demystified. The Second Law of Thermodynamics Reduced to*
Plain Common Sense. World Scientific, Singapore.
Ben-Naim, A. (2008), *A Farewell to Entropy: Statistical Thermodynamics Based on Information*. World Scientific, Singapore.

Ben-Naim, A. (2009), *An Informational-Theoretical Formulation of the Second Law of Thermodynamics*. J. Chem. Education, **86**, 99.

Ben-Naim, A. (2010), *Discover Entropy and the Second Law of Thermodynamics. A Playful Way of Discovering a Law of Nature*. World Scientific, Singapore.

Ben-Naim (2011a), *Molecular Theory of Water and Aqueous Solutions. Part II: The Role of Water in Protein Folding, Self-assembly and Molecular Recognition*. World Scientific, Singapore.

Ben-Naim, A. (2011b), *Entropy: Order or Information*. J. Chem. Education, **88**, 594.

Ben-Naim, A. (2012), *Entropy and the Second Law. Interpretation and Misss-Interpretationsss*. World Scientific, Singapore.

Ben-Naim, A. (2015a), *Information, Entropy, Life and the Universe. What we know and what we do not know*. World Scientific, Singapore.

Ben-Naim, A. (2015b), *Discover Probability. How to Use It, How to Avoid Misusing It, and How It Affects Every Aspect of Your Life*. World Scientific, Singapore.

Ben-Naim, A. (2016a), *The Briefest History of Time*. World Scientific, Singapore.

Ben-Naim, A. (2016b), *Entropy the Truth the Whole Truth and Nothing but the Truth*, World Scientific Publishing, Singapore

Ben-Naim, A. (2016c), *Myths and Verities in Protein Folding Theories*. World Scientific, Singapore.

Ben-Naim, A. and Casadei, D.(2017a), *Modern Thermodynamics*, World Scientific Publishing, Singapore.

Ben-Naim, A. (2017b), *Information Theory*, World Scientific Publishing, Singapore

Ben-Naim, A. (2017c), *The Four Laws that do not drive the Universe*. World Scientific Publishing, Singapore.

Ben-Naim, A. (2017d), *Entropy, Shannon's Measure of Information and Boltzmann's H-Theorem, in Entropy*, 19, 48-66, (2017)
Boltzmann, L. (1877), *Vienna Academy*. **42**, *"Gesammelte Werke"* p. 193.

Boltzmann, L. (1896), *Lectures on Gas Theory*. Translated by S.G. Brush, Dover, New York (1995).

Brillouin, L. (1962), *Science and Information Theory*. Academy Press, New York.

Brush, S. G. (1976), *The Kind of Motion We Call Heat. A History Of The Kinetic Theory of Gases In The 19th Century, Book 2: Statistical Physics and Irreversible Processes*. North-Holland Publishing Company.

Brush, S. G. (1983), *Statistical Physics and the Atomic Theory of Matter, from Boyle and Newton to Landau and Onsager*. Princeton University Press, Princeton.

Callen, H.B. (1960), *Thermodynamics*. John Wiley and Sons, New York.

Callen, H.B. (1985), *Thermodynamics and an Introduction to Thermostatics*. 2nd edition. Wiley, New York

Clausius, R. (1879). *The Mechanical Theory of Heat,* Macmillan & Co, London.

Clausius, R. (1865), Presentation to the Philosophical Society of Zurich.

Carroll, S. (2010), *From Eternity to Here, The Quest for the Ultimate Theory of Time*, Plume, USA

Cooper, L. N. (1968), *An Introduction to the Meaning and Structure of Physics*. Harper and Low, New York.

Cover, T. M. and Thomas, J. A. (1991), *Elements of Information Theory*. John Wiley and Sons,

New York.

Davies, P. (1995), *About Time; Einstein's Unfinished Revolution,* Viking, Great Britain

Davies, P. C. W. (1971), Space and Time in this Modern Universe, Cambridge University Press, Cambridge.

Davies, P. C. W. (1974), *The Physics of Time Asymmetry*, University of California Press, Berkeley.

Denbigh, K. (1981), *How Subjective id Entropy?* Chemistry in Britain, **17**, 168.

Denbigh, K.G. and Denbigh, J.S. (1985), *Entropy in Relation to Incomplete Knowledge*. Cambridge University Press, Cambridge.

Denbigh, K.G. (1989), *Note on Entropy, Disorder and Disorganization.* Brit. J. Phil. Sci. **40**, 323.

Eddington, A. (1928), *The Nature of the Physical World*. Cambridge University Press.

Gamov, G. (1940), *Mr. Tompkins in Wonderland*. Cambridge University Press, Cambridge.

Gamov, G. and Stannard, R. (1999), *The New World of Mr. Tompkins*. Cambridge University Press, Cambridge.

Gibbs, J.W. (1906), *Collected Scientific Papers of J. Willard Gibbs*. Longmans, Green New York.

Goldsein, S. (2001), *Boltzmann's Approach to Statistical Mechanics*. (Published in arXiv:cond-mat/0105242,v1, 11 May 2001).

Goldstein, H., Poole, C. P., and Safko, J. L. (2002), *Classical Mechanics* 3rd edition. Addison, Wesley, New York.

Greene, B. (2004), *The Fabric of the Cosmos, Space, Time, and the Texture of Reality*. Alfred A. Knopf.

de Groot, S. R., and Mazur, P. (1962), *Non-Equilibrium Thermodynamics*, North-Holland Publishing Co., Amsterdam

Guggenheim, E.A. (1949), *Statistical Basis of Thermodynamics*. Research, **2**, 450.

Hawking, S.W. (1988), *A Brief History of Time, From the Big Bang Theory to Black Holes*. Bantam Books, New York.

Hawking, S. and Mlodinov L. (2005), *A Briefer History of Time*, Bantam Dell, New York

Hill, T. L. (1960), *Introduction to Statistical Mechanics*, Addison-Wesley, Reading, Massachusetts.

Jaynes, E.T. (1965), *Gibbs vs Boltzmann Entropies*. American J. of Physics, **33**, 391.

Jaynes, E.T. (1957a), *Information Theory and Statistical Mechanics*, Phys. Rev., <u>106</u>, 620 Jaynes, E.T. (1957b), *Information Theory and Statistical Mechanics II*, Phys. Rev.,<u>108</u>, 171

Jaynes, E.T. (1973), *The well-posed Problem*, Chapter 8 in Jaynes (1983).

Jaynes, E.T. (1983), *Papers on Probability, Statistics, and Statistical Physics*,

Edited by R.D. Rosenkrantz, D. Reidel Publishing Co., London

Kafri, O. and Kafri H. (2013) *Entropy - God's Dice Game*. CreateSpace Independent Publishing Platform.

Katchalsky, A. (1963), *Non equilibrium thermodynamics*. International Science and Technology, **43**.

Katz, A. (1967), *Principles of Statistical Mechanics: The Informational Theory Approach*. W. H. Freeman, London.

Kondepudi, D. and Prigogine, I. (1998), *Modern Thermodynamics From Heat Engines to Dissipative Structures*, John Wiley and Sons, England Kondepudi, D. (2008), Introduction to Modern Thermodynamics, Wiley and Sons, Chichester, West Sussex, England

Kreuzer, H. J. (1981), *Non-equilibrium Thermodynamics and Statistical Foundations*, Oxford University Press,

Lebowitz, J.L. (1993), *Boltzmann's Entropy and Time's Arrow*. Physics Today, **46**, 32

Lebowitz, J.L. (1999), *Microscopic Origins of Irreversible Macroscopic Behavior*, Physica A,
263, 516.

Lemons, D. S. (2013), *A student's Guide to Entropy*. Cambridge University Press.

Lewis, G. N. (1930), *The Symmetry of Time in Physics*. Science, **71**, 569

Lindley, D. V. (1965), *Introduction to Probability and Statistics*. Cambridge Univ. Press, Cambridge.

Lloyd, S. (2006), *Programming The Universe, A Quantum Computer Scientist Takes On The Cosmos*, Alfred A Knopf, New York

Mackey, M. C. (1992), *Time's Arrow, The Origins of Thermodynamic Behavior*, Dover Publications, New York.

Muller, R.A. (2016), *Now, The Physics of Time,* W.W. Norton and Comp, New York

Penrose, R. (1989), *The Emperor's New Mind*. Oxford University Press, Oxford

Rifkin, J. (1980), *Entropy: A New World View*. Viking Adult.

Rushbrooke, G. S. (1949), *Introduction to Statistical Mechanics*. Clarendon Press, Oxford.

Sachs, R. G. (1987), The Physics of Time Reversal, The University of Chicago Press, Chicago

Sackur, O. (1911), *Annalen der Physik*. **36**, 958.

Scully, R. J. (2007), *The Demon and the Quantum. From the Pythagorean Mystics to Maxwell's*

Demon and Quantum Mystery, Wiley-VCH Verlag GmbH & Co. KGaA

Seife, C. (2006), *Decoding the Universe. How the Science of Information is Explaining Everything in the Cosmos, From our Brains to Black Holes*, Penguin Book, USA

Sethna, J.P. (2006), *Statistical Mechanics: Entropy, Order Parameters and Complexity*, Oxford University Press, Oxford

Shannon, C. E. (1948), *A Mathematical Theory of Communication*. Bell System Tech. J., **27**.

Sheehan, D.P. and Gross, D.H.E. (2006), *Extensitivity and the Thermodynamic Limit: Why Size Really does Matter*. Physica A, **370**, 461.

Sommerfeld, A. (1956), *Thermodynamics and Statistical Mechanics*. Academic Press, New York.

Styer, D.F. (2000), *Insight into Entropy. American* Journal of Physics, **68**, 1090.

Styer, D.F. (2008), *Entropy and Evolution*. Am. J. of Physics, **76**, 1031.

Tetrode, H. (1912), *Annalen der Physik*. **38**, 434.

Thomson, W. (1874), *Proceeding of the Royal Society of Edinburgh*. **8**, 325.

Tribus M. and McIrvine, E.C. (1971), *Entropy and Information*. Scientific American, **225**, 179.

Uffink, J. (2001), *Bluff Your Way in the Second Law of Thermodynamics,* Stud. Hist. Phil. Mod. Phys. 3, 305

Volkenstein, M. V. (2009), *Entropy and Information*. Birkhäuser, Berlin.

Wehrl, A. (1991), *The Many Faces of Entropy,* Reports on Mathematical Physics, 30, 119

Index

www.ingramcontent.com/pod-product-compliance
Lightning Source LLC
Chambersburg PA
CBHW030006190526
45157CB00014B/446